Visualizing Document Processing

Text, Translation, Computational Processing

6

Editors
Annely Rothkegel
John Laffling

Mouton de Gruyter
Berlin · New York

Visualizing Document Processing

Innovations in Communication Patterns and Textual Forms

by
Graziella Tonfoni
Lakhmi Jain

Mouton de Gruyter
Berlin · New York

Mouton de Gruyter (formerly Mouton, The Hague)
is a Division of Walter de Gruyter GmbH & Co. KG, Berlin.

♾ Printed on acid-free paper which falls within the guidelines
of the ANSI to ensure permanence and durability.

ISBN 3-11-017946-6

Bibliographic information published by Die Deutsche Bibliothek

Die Deutsche Bibliothek lists this publication in the Deutsche Nationalbibliografie; detailed bibliographic data is available in the Internet at <http://dnb.ddb.de>.

Foreword

"Information analysts are made, not born" might be the guiding idea of this book, which aims at the seemingly self-evident: handling text in order to cope with information. This idea is elaborated through a particular way of viewing language and text. The focus is not on bits or bytes even if information technology and/or knowledge representation in artificial intelligence are in the background, but on communication and its impact on the modeling of information and text processing, both for human use as well as for technical devices.

The crucial issue is the handling of text that is considered to be the linguistic realization of information management. Even if the approach is oriented towards cognition and information technology, the focus is on the integration of information into human communication. Text is understood as a vehicle for communication which allows the transfer of information, depending on specified communicative purposes and intentions. How and why this can happen is represented through a set of text procedures, for which a system of visual representations is developed. They are represented in terms of graphical units composed according to principles of painting and architecture or according to an intuitive understanding of pictures. There are both simple and compound signs and symbols which operate as a kind of notation and which, as a whole, provide a complete metalanguage (representation language) for specifying the different text procedures. These procedures are meant to serve both for analysis/ interpretation of texts as well as for text construction in terms of information handling.

The focus is on the core grammar of the communicative competence in terms of articulation of the dynamic aspects of text processing. The principle of correctness in syntax is paralleled through the principle of consistency, which refers both to the matching of the communicative text representation levels and to the cognitive program according to which this match is realized.The conceptual tools which are visualized are grouped as textual canvases, textual machines which might be related, textual objects and a set of icons (signs and symbols):

Canvases are considered to provide the space of an abstract texture (pattern), which is progressively filled with regard to a set of aspects such as density, networking and chaining words, metaphor creation, pattern recognition, narrative text planning, story planning, text drawing, dialogue, collage writing and mosaic reading in teams, text rotation and hypertext.

Machines represent the different cognitive processes in the view of a dynamic three-dimensional visualisation tools. There are processes and machines such as simple and complex rotation of a word (associating words and chaining them), the text motion machine that develops the stories in a narrative text, but leaves them open, a textual lens that focuses on some part of the text and expands that aspect which is most relevant. Other examples are only enumerated: wheel for text browsing, carriage for memory text, carriage for explanatory text, three-floor rotation machine for poetry, machine for transferring text, for transforming text in phases, focusing machine, machine for producing ideas and turning them to text, and the textual catapult.

Signs control the specific communicative function of an unit such as square (general information), square within a square (reduced information), frame (a story is analoguous to another story), triangle (memory text and subjective experience triggered by previously provided information), circle and grouped semicircles (global concept(s) emerging from a previously provided general information or story), semicircle (particular concept extracted from previuosly provided general information), inscribed arcs (adding an alternative to a story or general information), opened text space (opening a new text segment), right triangle (comment deriving from previuosly provided general information).

Symbols refer to the communicative intention at the local level. Examples are *decribe*, *define*, *narrate*, *point out*, *explain*, *regress*, *inform*, *reformulate* and *express*. Turn-taking symbols assign a specific role to the reader with regard to his/her interpretation. Examples are: major and minor scale indicate that the text should be interpreted as it is or as a modification; open and tight rhythm provide information about whether the text is complete or in need of completion.

Textual objects refer to the construction of the text: cube (shape organization for constructing a story), pyramid (personal memories about a story), miniature (organization for expanding and completing texts), polittico (discussing a topic), obelisk (construction of a common story by a group), cylinder/column/papyrus (explanatory text construction).

The theoretical foundations consist of a newly developed theory of language (and text) which, on the one hand, is meant to correspond with some intuitive knowledge of speakers and writers and, on the other, is applicable to practical text (or document) management tasks. This claim is part of the scientific credo which is continuously emphasized in the book.

As a consequence, the method chosen for the development of the system is based on the assumption that a pictorial approach is the best way of detecting and conveying knowledge about the non-visible. This is the reason why metaphors play such an important role. It also reveals the AI roots of thinking and modeling, where metaphors and analogies are used in order to articulate and bring to the surface abstract information processes which are not visible and easily accessible to understanding. In this view, Minsky's "society of mind" (1986) is the guiding model for this method of system development. Metaphors, especially with respect to their conceptual foundations in the various source domains, provide the system of terms and concepts for the proposed theory of language. As language is considered to be a "natural phenomenon", the natural sciences, mainly physics including mechanics, optics, quantum theory, nuclear physics, but also chemistry, are selected as "source domains".

The book does not only deal with the construction of the system as such, but also with the kinds of methods that are to be applied to this construction and, more generally, with the way in which this approach fits into scientific activity. Thus, a framework for the specific theory of language is provided, linking the present approach to innovations such as were introduced through Chomsky and Searle into linguistics or through Plank and Einstein into physics.

Operationality is satisfied through the application of the visual representation system as an annotation language (mark-up language) for document retrieval. The visual system makes it possible to track the various changes in information during document construction. Thus, it can provide a trace of the history of a document in order to keep this information accessible and reusable for later purposes if contexts have changed.

The argumentative thread running through the book is as follows:
The focus is on the representation of text procedures in terms of definitions and their visualizations (Chapter 3). The understanding of these constructions is prepared by the introduction of the conceptual system and the dicussion of the various scientific paradigms which have an impact on them (Chapter 1) and the development of a particular

theory of language and text (Chapter 2). The representation of the system itself is followed by a dicussion of the application of the visual system as a kind of mark-up language (annotation language for documents) and its role in the work of information analysts (Chapter 4). The theoretical framework which starts with Chapters 1 and 2 is completed by a final chapter. The crucial role of metaphors and analogies as tools for scientific exploration receives additional emphasis in Chapter 5.

According to the specific style of research theory and system are developed in terms of definitions, analogies and metaphors, with emphasis being given to evidence provision and justification. In addition to this, the role of the researcher and his/her responsibilities in society are discusssed. This includes the choice of questions that are investigated as well as the manner in which results are conveyed to the members of society. The writing is characterised by a high degree of accuracy and explicitness that might be in conflict with the readability of the text. The world of the author is not immediately accessible to the reader. Another point which might have an impact on reading is that the various aspects of the system are discussed from different point of views. Thus, although the reader may be confronted with relatively many repetitions, s/he is also allowed non-linear access to the content.

The scientific relevance of this approach fits very well the profile of the TTCP-series. What is provided is a closed system of information and text processing that integrates different points of views from linguistics, cognition, communication theory, artificial intelligence and everyday knowledge. This interdisciplinary approach is grounded in a kind of 'cognitive pragmatics' in the perspective of linguistics providing a model of the 'competence of performance' (on analogy with Chomsky's syntactic approach).

Even if the importance of some particular metaphors of the theory of language might be more or less open, as a whole they are fascinating and stimulating.The relevance of the text procedures as a complete representation system is beyond question. The basic ideas of this system are unique; the consistent elaboration in terms of the visual representations is both theoretically interesting and practically relevant. This level of integration of theory and practice is rare in the disciplines concerned.

The volume is the condensed result of research carried on by Tonfoni and knowledge-engineered by Jain. This innovative fusion provides a closed system of information and text processing that integrates different points from linguistics, cognition, communication theory, artificial intelligence and knowledge management.

The Series Editors, 2004

John Laffling
Edinburgh

Annely Rothkegel
Hannover

Preface

This research-oriented book is intended to be for a specific audience, represented mainly, but not exclusively by information specialists whose work is based on document reception and production. Accuracy and reliability are for them not an option, but an imperative. They need therefore to be provided with special and particularly sharp conceptual tools, theories and methodologies so as to be able to grasp the complexity of communication and information, which has been exponentially increased today by the pervasive and massive use of information technology and the huge amount of data available. The book may be also seen as a kind of systematic and scientifically based justification of a new way of thinking about information design and information analysis.

In the first chapter of this book, the readers will find the theoretical background of a new approach toward language and information. The new terminology is introduced and the main concepts are defined. We have used interdisciplinary approach in defining scientific and linguistic paradigms. The main paradigms for incorporating intelligence in machines are presented. These include knowledge-based systems, artificial neural networks, evolutionary computing, fuzzy logic and artificial agents.

In the second chapter readers will be led to the implications resulting from rethinking of language and text through the illustration of both the theory of text comprehension and of text compression.

The third chapter illustrates in a more detailed way a new way of looking at communication by familiarizing the reader with a highly articulated visual system for interpretation of qualitatively different components in natural language. As for the analysis of physical matter, many levels of observation may be considered and legitimately adopted, in order to focus on different aspects both at a macroscopic and at a microscopic level. In the same way, new tools for observing, describing and explaining qualitatively different phenomena in natural language are introduced here.

The fourth chapter illustrates most recent evolutions of the model provided and its final output: CTML, a powerful mark up language used for strategic documentation annotation.

The fifth chapter contains a set of theoretical reflections about what metaphor making in the science of information today may require and implies, along with practical suggestions for verifying and augmenting the consistency and relevance of analogical reasoning.

We are pleased to acknowledge the help of many people and various sources in the preparation of this book. Some of these include:

- The *Encyclopedia Britannica,* which has provided the definitions of many of the currently used terms in physics, chemistry and biology.
- The *Random House Dictionary* has enabled to check definitions of those terms.
- Annely Rothkegel, John Laffling and anonymous reviewers for their insightful comments which have improved the quality of this book.
- Feng-Hsing Wang and the staff of the Walter De Gruyter & Co. including Birgit Sievert and Wolfgang Konwitschny for their assistance in the preparation of the manuscript.

Graziella Tonfoni, Italy and Lakhmi Jain, Australia
2004

Table of Contents

Chapter 1

Looking at language and information

1.1 Introduction

This chapter presents a new theoretical framework of qualitative reasoning about communication. This is required in order to understand why any natural language processing system must be grounded on a consistent and scientifically based natural model. Such a new framework of qualitative reasoning is primarily concerned with aspects of communication, which may be defined as qualitative aspects of communication and information occurring in natural language.

In order to identify all of these aspects, a scientific framework is formulated to handle "facts" of language. These need a systematic theory for their explanation, and not merely be described. Any language scientist needs to be interested not merely in describing and cataloguing facts but also in creating a consistent theory or set of theories that may explain and justify such facts. Any philosopher in the science of language, has to analyze isolated language facts and also analyze the way those facts have been previously identified and consistently named. Language scientists may define specific problems to be solved only by creating a whole set of consistent rules and criteria.

1.2 Empirical and conceptual foundations

A basic task of the science of language is to explain language facts and communicative events taking place "via natural language".

Language facts as well as communicative events may be discovered through observational methods. Listening to a conversation when it occurs naturally and recording it, or reading a text and analyzing it, without trying to influence what is happening, may be turned into a kind of experiment aimed toward analyzing occurring events so that some consistent model of interpretation may be generated retrospectively.

Chomsky (1957, 1965) has provided language scientists with a model that promotes a whole set of "ad hoc" experiments, so as to observe language facts by asking for evaluation and judgment about the phenomena to the native speaker.

In the present observation model there is no predesigned experimental situation. This ensures that the individual under observation is acting spontaneously. There are however strong and stable points of reference for observation and evaluation on the part of the observer.

There are many examples in science, where processes and phenomena that cannot be seen may be discovered by using experimental methods employing special equipment, which enables the phenomena to be observed. 'Special equipment' in my observation procedure resides in the capacity for self-observation and self-reflection. The observer can recognize what is happening to individuals observed so that they may provide a system of reference for identifying and naming the observations. This can form the basis of a system of reference, which provides new definitions and a specific terminology. Without a shared system for interpretation, there may be no observation of facts, just random data collection.

Every science employs its own abstractions, terminology and techniques for interpretation and explanation of facts. These may actually be substantially different. Conceptual tools are ways to make phenomena intelligible. In the present approach these conceptual tools are used to detect linguistic phenomena of relevance. There are many such visual tools.

1.3 Relation of theory to observation

In creating a theory we may distinguish between an empiricist attitude and a rationalist attitude. For empiricists, raw facts of experience are primary, and theoretical principles are accepted only when interpreted as generalizations of empirical data. Theoretical entities have to be understood as derived and constructed from empirical experiences and need to be strictly related to them.

For rationalists, the scientist's experience is just a disorganized aggregation until some structure and principles are found. A scientist needs therefore to be selective in observation. Well designed experiments are viewed as "valuable ways to create phenomena which may be significant" (Encyclopedia Britannica).

This theoretical approach intends to combine harmoniously both perspectives: on the one hand, there is actual observation of facts derived out of experience, and then interpreted to derive valid laws. On the other hand, there is a consistently structured framework providing specific criteria and points of reference.

As any theory requires, there are empirical foundations and a strong and consistent internal organization. This is so defined to ensure that the parameters chosen are reliable and consistent.

1.4 A 'new' framework

One of the basic issues in the philosophy of science is the identification of different possible styles of explanation in the various sciences, at different stages and of different stages in a given science. One main concern is also "recognizing periods of shifts in interpretation of concepts and building up of new concepts" (Encyclopedia Britannica).

The present approach has evolved progressively over time and has been extended in many directions. Firstly it was based on direct investigation. Working on a theory that could explain and justify facts and events, and generate new processes, which could not even have been thought of in the old terms. The question was, which kind of task the new explanatory framework in the science of language could perform and which requests it could satisfy.

What became evident in the evolution phase was that a consistent environment for observing invisible phenomena and highly abstract processes evolved. It was found by this means I had an entirely new kind of scientific approach to language. This resembles research in natural sciences, such as botany or entomology, and formed a sort of physics of language. It identified basic components of language matter and basic energies and forces. These may be thought of as active. Furthermore the new language components capable of undergoing special kinds of transformations were discovered, which I named "reactions". This is similar to the reactions in chemistry. It has resulted in an awareness of an important distinction existing between language facts, communicative events and the possibility of recreating and reproducing communicative chains of reactions. The new framework for analyzing language resulted into something completely different, which could have applications in real life communicative environments.

1.5 A new way of thinking about language: from texture to text

This approach and exploration of phenomena is new for the following reasons:

1. Using highly specific metaphors it has been possible to create a common metalanguage that can be shared by the scientific community, to represent text operations. These may become precisely defined communicative actions, which may easily be used for text processing.
2. Thinking of text as the final result of a highly complex and dynamic set of processes implies that its "structure" may have previously undergone many changes. Text may be therefore defined as the result of a whole set of text actualization processes, derived from an overall basic structure modified at different times.
3. At the "structure" or "texture" level it is possible to see and predict all the dynamics and changes happening before it is actualized into a real text.
4. A text may be seen as a really dynamic structure and a model may be set up which derives from mechanics. Mechanics is that part of physics that states the specific conditions of movement and equilibrium. In other words, the intention was to capture and represent those dynamic processes that we may observe to be active in text, by a mechanical framework.

This theory of language may sound like physics which is the study of natural phenomena, or chemistry, which is the science that analyzes the composition and structure of elements and explains transformations. Language may be thought of in these ways, as it requires observation of the current communicative actions and is subject to continuously occurring transformations.

1.6 Text representation systems

To explain in a detailed way different phenomena occurring such as the interplay of communicative function, intention and turn-taking within an actual text, a set of representation systems are designed. These are named as metaphors and analogical reasoning textual canvases, textual machines, textual objects, textual signs, textual symbols and textual turn taking symbols.

This part of the theory may be considered as a kind of textual optics where basic conditions of textual vision and textual perception may be illustrated and thoroughly explained.

Some further work to compress and condense textual information further may be thought of as a kind of kinetics, as it describes the conditions for dynamic conditions and a new "movement" in textual interpretation. This is actually acceleration and transformation of single sentences and paragraphs of a document. Parts of an already interpreted and well understood document may in fact undergo reformulation processes according to a set of priorities changing over time.

1.7 A quantum theory of language

Up to the present time linguists have never been concerned with problems of acceleration of text segments. Transformations have been considered and defined by Chomsky (1957) as acting on a very different level which is the syntactical one.

Looking at what was already available in the field of linguistics, it is clear that Chomsky created an extremely relevant, "corpuscular theory of language", and Searle (1968) introduced the idea of an "undulatory approach to language". It is proposed there is a third way, which may complement the earlier ones. This is an analogous development similar to that which Max Plank did for physics when he stated that energy is both discrete and complex with no resulting contradiction.

We may therefore think of a quantum theory of language based on elementary and complex textual units, carrying communicative energy of three kinds:

1. communicative function,
2. communicative intention,
3. communicative turn-taking.

Einstein underlined the relevance of Plank's theory in 1905 by suggesting that a quantum could be considered as a "package" of light radiation.

A compressed textual unit may be equally well thought of as a package of linguistic information charged with communicative energy, which may either be retained or transformed. It is both an elementary unit and a complex one, just as the quantum is in physics. In order to avoid the dichotomy between "discrete" and "complex"; a compressed textual unit, which means a simplified and compacted sentence, is both discrete and complex.

We may also draw a fundamental distinction between a theory of comprehension and a theory of compression: even though they both refer to text, they operate at very different levels. The theory of text comprehension applies to a potential text, and is concerned with the possibility of actualizing all of the texture potential into as many texts as possible in the forms of various editing outputs. The theory of text compression applies specifically to compressed textual units, which have already been processed and checked and made fully compact.

Both theories are strictly linked together: the theory of compression is the logical consequence of the theory of comprehension. In other words, a compressed unit is the result of an actualized text, which has been processed and consistently represented according to the overall system of signs, symbols and other visual schemes, which are part of the theory of comprehension. No text can be consistently compressed if it is not first interpreted according to a consistent system of qualitative reasoning about communication.

1.8 A qualitative communication framework

Within a newly established qualitative communication framework, we may be primarily concerned with identifying, defining and consistently applying qualitative reasoning on communicative actions which manifest themselves through sentences and paragraphs within a text. In order to state basic concepts in qualitative reasoning about communication occurring through natural language, a consistent set of visual tools specifically aimed toward representing invisible aspects of natural language "by turning them into visible and tangible elements" is very useful. Each of these conceptual tools for explaining language phenomena, may also explain new language effects and reactions, just as in physics and chemistry.

The possibility of controlling transformations and reactions occurring within a certain text undergoing different kinds of reformulation phases, opens up incredible opportunities. As with any discovery or invention, it is extremely important to explore such possibilities.

This theory of language, as well as the information control environment, are aimed toward enhancing transparency and effectiveness in communication and information processing by first promoting awareness and self-reflection processes.

1.9 The 'user' becomes 'chooser'

Being able to see, identify and name processes occurring in information means becoming active protagonists of a crucial aspect of life in the era of information technology and information workload as well. According to the framework here presented, the "users" of a certain technology are turned into "choosers", not only because they need to be free to choose whether to use the new tool or not, but also because they will have to become most sensitive to the inner nature of information. In order to become really sensitive, they will have to become aware of the implications deriving from the various ways of processing information.

There is a basic capability for communicating and organizing information as whenever we use language we communicate and organize information. There is a need today to improve ordinary skills and promote a higher communicative perception. Looking at things and observing them is different to looking at things through a camera as it is to take a picture. The important thing is that there is a different way of seeing, not merely the camera itself.

As a result of looking at things differently while using a camera, we get a picture, which is different from what we thought we could see by using our eyes.

1.10 What if a camera was a human eye?

In Tonfoni's theory of text comprehension, it was necessary to draw a distinction between texture and text. Texture is the result of a continuing flow of sentences, occurring naturally, when we talk and write, listen and read. Text is the result of consolidated sequences, which show patterns.

The fundamental difference between human sight and seeing through a camera is that the human eye sees the picture in its context, whereas the same human eye may see the same object differently through a camera in order to turn it into a picture. This is because the context may be missing.

This is similar to the difference which exists between texture and text. Text is like a picture, whereas texture is like sight. A picture has to be independent from its visual context, just as the text has to be independent from its communicative context. Nevertheless both picture and text retain those distinctive elements, which are part of their own context. A picture may show differences in distance, perspective, color reproduction and movement: all of these differences go back to the basic distinction between human sight and what a camera enables human sight to see. In the same way a text may show differences in perspective, here called communicative function, in color reproduction, here called communicative intention and in distance and movement, here called communicative turn-taking.

Just as there is a precise way of looking at things determined by a camera, in the same way different kinds of perception may occur when reading and interpreting a text. A theory of text comprehension may well supply the information analyst with a whole set of tools to regulate and reproduce originating texture elements that may be lost in text, very much like a photographer who uses special techniques and tools to reproduce and exaggerate something that would otherwise be lost.

1.11 The textual pentagram

A theory of text comprehension will lead an information analyst toward conceiving a document as a textual pentagram with a clear and precise system of textual notation complementing text. Textual notation, just like musical notation, may have a precise value and may consist of a set of highly articulated interpretative clues that are consensually shared and agreed upon.

What will be here illustrated is a specific system for interpretation, meant to allow an information analyst to identify and describe the different kinds of operations, which are likely to surface out of a given document.

As such, it may productively be used as a consistent system of instructions for indicating what has happened in producing a certain text, but it may also be used as a reliable device for changing a text according to very specific indications which may be perceived and understood as consistently represented and consensually agreed upon.

1.12 Nuclear energy in text processing

The theory of text comprehension allows some 'X-raying' of information. In other words, it permits one to see "behind and beyond just words, paragraphs and texts". The theory of text compression allows communicative energy to be transformed and information reactions to take place within a certain text, under certain circumstances and within a specific information processing environment. The details of all the possible foreseeable applications have been tested as accurately as possible. It is hoped that these discoveries will be freely available and used for the good of the community.

1.13 The scientist needs to feel responsible for both theories and applications

A responsible scientist, while making important discoveries, should never forget to pay attention to educational environments. Supporting teaching with experiments will favour experimental reasoning by the students. A text comprehension theory may be effectively applied to a large set of different domains.

Provided they are subject to control, a text compression theory, derived as a natural consequence of the text comprehension theory, may be used for precise and useful applications. Analyzing sets of documents, which may result from a text compression process, is a delicate issue. Here the control within the framework resides both in the encoding and in the decoding processes. Each compressed information unit may be retrieved and recognized in its original state.

Any transformation process, according to this system, leaves a trace which guarantees that all operations performed upon a certain document may be traced back and recognized. Tonfoni has followed the same principle and leaves visible all that has happened in text-acceleration and when text reactions occur. The crucial matter is where the control and transparency function will be active and up to which level.

Time and energy have been dedicated to the design of educational applications in order to promote cognitive awareness. This support of cognitive self-reflection tools means that individuals are more in control of the information processing activities and communicative actions. Self-reflection is a process which allows individuals to see and understand what is happening in communication. They are able to recognize communicative patterns emerging out of each document under analysis.

1.14 Self-observation as a scientifically supported procedure

In science the very idea of self-observation is a delicate matter. Natural sciences have always used observation, and increasingly accurate tools have become available for observations. This equipment enables researchers to see what could not be seen by the naked eye.

Visual representation and a highly consistent terminology are used to communicate the result of such observation. In linguistics observation is used as a means for categorizing languages and identifying different various kinds of communicative behaviour. Chomsky sees observation as a way for the linguist to gather and interpret incoming data according to a certain theory. The chomskian linguist obviously triggers syntactic self-observation just by asking native speakers certain kinds of questions. The questions cause them to reflect and give opinions about different syntactic actualizations.

For this system experts will at first make their own observations and then extend the observation process to the other individuals. It is clear that communicative patterns recognition has nothing to do with psychoanalytical search based on criteria, which are of a fully different nature.

1.15 Some important previous assumptions

At this stage it is necessary to provide readers, such as researchers in the fields of linguistics, artificial intelligence, cognitive science and information design with some more information on the new models here introduced. The first step is to give recognition to the basic fact that language is an extraordinary complex issue. To handle such complexity, it was realized that existing models could only serve as a point of departure for a very long journey.

It soon became clear that one theory could only take into account some of the processes and phenomena but others would be left out. Those aspects of language, which were excluded, could then be described, defined and explained within a new framework, which would serve as an extension of the previous framework. Each framework would then be complete in itself but would still need to coordinate with the others according to some progressive order. In a similar way to the distinction drawn between texture processes occurring before any real text may take place.

As in an engine, differing interpretative devices activate and regulate different parts, which are all compatible, consistent and synchronously working. Behind each of the identified devices in the theory of text comprehension is a precisely reframed model of reference.

Only when metaphors may sound clear but they may turn into real conceptual tools precisely defined as to explain various kinds of processes. To make sure no confusion arises, what may be defined as a "process" is in fact a dynamic set of cognitive operations resulting into linguistic facts and events and communicative actions. Phenomena are the output result of processes and therefore refer precisely to linguistic facts, events and communicative actions.

1.16 Some very important background

By closely exploring Chomsky (1957, 1965), it was evident that a set of basic and important assumptions were still extremely relevant. These are as follows:

1. The need to state generally valid laws in language.
2. The need to set a scientifically based framework to study language phenomena, proceeding from observation and moving toward both description and explanation.
3. The need to represent visually dynamic aspects of language such as generation and transformation throughout an adequate, both consistent and consensual system.

4. The need to identify and name newly discovered and invisible transformations and movements, which work on each single language they apply to.
5. The need to state a consistent background so as to allow each individual to access a common system, both independently and compatibly with cultural background diversity and single language differences.
6. The need to state different levels such as a deep structure and a surface structure so as to justify different kinds of transformations occurring.

According to the new approach proposed here, the focus is not on syntax, nor is the sentence the basic unit to work on. Syntax comes last and language is analyzed within a more general framework aimed toward processing linguistic sequences and those invisible elements that are so important. If these invisible elements are not taken into account properly, they deeply compromise any kind of linguistic performance.

1.17 On redefining 'text'

Text is the basic entity of this research. Text in this case means precisely "the basic communicative complete entity". In the course of developing a theory of text comprehension, further specifications are introduced as required. These clearly discriminate between text and texture, text and text segments and also introducing global control text instructions and local control text execution cues. By carefully exploring the artificial intelligence tradition in knowledge representation and in vision, the basic assumptions are followed as:

Any theory will prove to be valid if and only if it can be tested and be used practically. In other words, not only is it important for a theory to be based on correct and consistent assumptions to prove its validity but it needs to be operationally valid. That is able to support consistent models, which can justify very specific aspects of linguistic performance.

1.18 Consistency in new terminology

This theoretical and methodological approach is called the communicative positioning program (CPP). In this approach, text and sentences represent, linguistically, a precise set of communicative functions, communicative intentions and communicative turn-taking attitudes.

This kind of information is conveyed not only directly by words, sentences or other linguistic elements, but also comes from an "on top" interpretation model. These invisible aspects of language are named as the "communicative positioning" of a certain text, which reflects the text producer's communicative attitude toward the receiver of text. Firstly, many of those common naturally occurring interactions are observed, and they take place in different languages and in various cultural contexts. An extensive reading of a variety of languages and cultures was undertaken. A wide variety of different topics was read to ensure that the set of recursive patterns obtained was diverse enough to be declared generally valid. Finally, all of these invisible components were defined in a way to ensure there were just enough and that they would cover the whole variety of possible cases occurring. In addition they could be combined so as to cover any kind of occurrence. Since these elements were originally invisible, it was decided to make them visible so that they could be identified and named consistently. This is the reason why a visual representation system was created for each of them. Each element was both natural and intuitively understandable, once illustrated. Finally, a Text Representation System (TRS) was defined. This is a visual set of conceptual tools, which represent qualitatively different phenomena occurring in text. A quantum theory of language was evolved to address the problem of qualitative reasoning about language first. This was the next step after the text was recognized, analyzed and defined. The qualitatively different phenomena occurring in language performance was then described.

1.19 A productive distinction: deep versus surface

A productive difference between a deep structure of texture and a surface structure of text allow us to distinguish between macro-level phenomena active at the deep structure level of texture – not yet linear text organization – and micro-level phenomena active at the surface structure of text-linearized text organization. The intrinsically dynamic nature of macro-level active phenomena is represented with visual dynamic representations such as textual machines and textual canvases.

Actually, both textual canvases and textual machines are intended to represent visually and dynamically very general processes and global phenomena that are difficult to grasp in text, because they occur at the texture level and only leave traces at the text level of what had indeed previously happened at the texture level. Such global processes do nevertheless play a crucial role, because without them taking place, no surface text linear organization could even be conceived. It can be stated that what is left at the surface structure of text, which is already the linear organization of text in sentences and paragraphs, is the result of deep texture phenomena having occurred and having left traces.

To ensure that those traces of deep texture phenomena already identified, defined and visually and dynamically represented, are correctly recognized, they were named according to their qualitatively different nature. It was realized that such traces are different in nature and therefore need to be represented in a way that will make them look like what they really are: different but acting together at the surface structure of text. By carefully exploring the nature of such traces, one can also see very clearly that they will work in a way that can be described in terms of energy.

1.20 Active energies in text processing

Within the newly established framework, let us envision some globally active energies and some locally active energies. There is no contradiction or contrast, just compatible action at different levels.

The issue at this point was for me to define them and represent them visually in a way that would allow them to be seen working together, both globally and locally, as execution cues and instructions for performance, which is consistent understanding of that sentence or text.

According to these specific and different aspects, let us make a preliminary distinction among communicative function, communicative intention and communicative turn-taking. These are precisely those invisible aspects of textual meaning that need to be recognized to guarantee an adequate performance.

Altogether they represent the qualitative aspect of a text and of its text segments. In order to be able to represent global indicators for text execution, carrying communicative function, many different languages and communicative contexts were checked as to make sure all the consistent and generally valid elements had been identified. Let us now move towards a specific definition of each of them, attributing highly consistent terms to each of them and finally choose a set of signs to indicate them.

1.21 The 'program'

To describe with a metaphor the visual syntax for text performance, the signs and symbols system may be easily defined as a communicative traffic control system and as a textual notation system, where signs represent global indicators and symbols represent local execution modes. As a consistent system in itself, signs and symbols may also work as a metalanguage, which may operate on top of natural language and may be considered as a programming language on its own, based on a precisely defined set of instructions native speakers should be able to recognize and apply to their own language.

It is important at this point for me to go back once again to the word "program", which was deliberately chosen. Program indicates different and highly compatible aspects in this theory, more precisely:

1. The cognitively consistent background for recognizing communicative positioning attitudes and for performing accordingly.
2. The set of communicative actions each individual may recognize to be represented in text.
3. The metalanguage that operates on top of natural language, therefore constituting a programming language by itself.

What has been described up to now largely refers to the theory of text comprehension, which was further developed and extended as to generate a consistent theory of text compression.

1.22 A theory of language must be able to work

A specific aim of this research has been to generate a theory of language, which could prove its validity and practical use. In other words, the idea was the linguist not just to observe or comment on language facts, but also to be able to build communication tools and be part of the construction of new communicative environments. In order to do so, linguists need to share a common vocabulary, to master their own tools, just like architects do. Architects plan first, then discuss the details of the project, then they may take turn in what they are building, just because they may talk very specifically about what they are doing.

The highly technical and precise meaning of the terminology will enable linguists to define the kind of operations they either perform or see performed through language. Quite a lot of undesired ambiguity may then disappear, as linguists can systematically define the kind of operations they wish to perform very quickly.

It could be a tremendously powerful example for other fields as well, if the scientific community of linguists could reorganize terminology in linguistics by using criteria that could then be transferred into other research areas, as well as into the education and business fields. A consistent way of learning what is going on behind and beyond words and sentences may be used within any field of knowledge.

1.23 On cognitive self-reflection

Self reflection is a very important ability, which needs to be acquired as we are not taught to do it though we may easily learn how to do it. Cognitive self-reflection is first of all a process of self-observation and careful discrimination among the kinds of cognitive actions, which we activate and become involved, and finally result into identifiable communicative actions, which is the way we talk and write.

As to make this process possible, those same attitudes which may be named in terms of communicative positioning may well support the individual during the observation process by constituting mutually shared and agreed upon interpretation cues. Those cues may be applied to a wide and various range of phenomena, according to a consistent, cognitively based and therefore generally valid terminology and syntax of language performance.

Just like language, syntax needs to be viewed as a stable system, which allows some combination and does not allow other ones. As such it may guarantee for correctness; nevertheless it is also meant to generate an infinite set of possible actualizations, in the same way a basic syntax for language performance is a guarantee for correct generation of sentences and for consistent interpretation of an unlimited set of texts.

Just as a syntactically correct performance may be acquired easily because of the existing basis of a syntactic competence (Chomsky 1957, 1965), in the same way communicative positioning attitudes may be conceived as a consistent result of a cognitively based set of communicative attitudes, which may first be recognized as active continually in currently occurring communicative performance and currently occurring communicative intercourses.

Referring back again to Chomsky (1957, 1965), we know that a native speaker may be able to judge and evaluate the validity of a certain sentence according to a uniform set of rules, which allow an individual to produce correct sentences. Those very same rules constitute a guarantee for both generation and evaluation of syntactically correct sentences.

Just as native speakers are able to tell what is syntactically acceptable and what is not, the very same way they need to be able to recognize the communication positioning they adopted and therefore reflects on their own text. Consistent recognition may be possible through some kind of intuition and sensitivity or may be systematically defined as resulting from a highly articulated process of self-observation. This is exactly meant by cognitive self-reflection. This may also extend to collective self-reflection, which entails precisely the interpretation of a complex set of communicative activities performed by many individuals, all together, then being interpreted in their own variety, by using a common system of reference for interpretation.

1.24 A highly reliable system for operating text

An intrinsic advantage of a commonly shared system of reference for interpretation of communicative activities is the possibility of actually having a solid basis for accurate planning as for very specific communicative actions to be undertaken. In other words a stable system will allow individuals not only to ask about some topic but also to specify the quantity and quality of information they may expect to be provided. This is possible only if the qualitative aspect of information has been recognized as playing a fundamental role and has therefore been declared relevant so as to be first analyzed, and then defined and visually represented. Once able to identify and define the qualitative value of a given text or part of a text, individuals involved with a certain communicative intercourse will not only be able to ask for a different quantity of information, but ask also for a different quality of information as for a certain topic, by defining precisely the kind of information they intend to get, being able to identify the desired qualitative value of the requested information.

By the same principle, they may then also modify, add, delete or substitute parts of a text by explicitly declaring the kinds of operations they have been doing. And any change may be made transparent thanks to a visual, consistent system of representation which may allow traceing back both the kind of choices made and the procedures used. Disagreements or misunderstandings which may occur will be dealt with and solved according to the same principle: they will first be named, then made transparent, then finally handled.

This approach may look simple after all. A system – which comes as the final output of a highly complex and articulated theory – is simple because it represents the end of much complexity that has been previously recognized, dealt with and organized.

Even using the terminology and those metaphors selected and redefined, it is very clear that many different approaches are involved to get to a final synthesis. Since the aim of the theory of text comprehension is to come up with a wide description, with a consistent definition and an appropriate explanation of text and text segments as communicative entities, many metaphorical tools have been presented in order to provide description and definition and then finally explanation too. Complexity inherent in natural language immediately raises the issue of compatibility and consistency of different models involved and ways of reasoning about linguistic phenomena.

It has been necessary to demonstrate how a certain framework, which had already been organized to provide description, definition and explanation for a certain set of processes and phenomena, could profitably coexist with a different framework, which could account for the description, definition and explanation of other processes and phenomena, which could not have been handled into account in any of the previous ones.

1.25 Bridging the gap between art and science

In order to be able to address the issue of describing, defining and explaining the complexity of communicative events, a new and highly articulated "metaphorical realm" is established. Not only, it is important for scientists to create new theories; they must also produce consistent explanations; neither is it enough to prove such theories work. Each scientist must care about the application of theories and the extent to which they can be positively applied to enhance society and individual growth. This is what science should be about: the construction and sharing, or discovery and access, to only that kind of knowledge that will be at no risk, through accurate design and testing of a consistent technology.

As technology is able to affect the "art" of doing things, by opening up new possibilities that trigger new attitudes, if well planned. To conclude: the intention has been to create a theoretical framework, which may support operational models and provide new tools, which will enhance responsibility and awareness in the communication and information processing era.

1.26 Biologically inspired information processing

Scientists and engineers are directing their efforts pertaining to the information processing problems towards techniques that rely on biological systems for inspiration (Jain 1995, 2001). As computers have grown in power and reduced in cost, there has been a thrust to apply them to automate an ever increasing range of activities. The primary reason for this thrust has generally been economic, although safety or reliability may also be considered. There may be a variety of driving forces, but it is clear that the trend to replace human being in an increasing range of activities is inexorable. The end product to which the new biologically inspired techniques aspire, either by design or implicitly, is a system that is capable of performing 'like a human'.

The end product, which the new engineering techniques aspire, either by design or implicitly, is a system that is capable of performing 'like a human'. It must have the ability to learn to perform tasks and make decisions, and in many applications it must have the flexibility to generalize 'sensibly' if it finds a situation for which it has not been trained. These attributes are generally the direct antithesis of the traditions of engineering or the applied sciences, where the historical basis for the disciplines has been the certainty and predictability of physical systems. The new engineering calls for techniques that go beyond conventional engineering and operate in varying environments, dealing with concepts that humans may describe in terms such as 'my best guess is' or 'common sense tells me'.

The limitations of conventional processing techniques in meeting the goal of human-like capability is very well demonstrated by the lack of progress in image processing. For years, frustrated researchers have worked on image processing tasks in the knowledge that the very best of their programs could not approach the capability of their own brain in performing even very simple tasks, such as finding objects in an image.

In order to develop new technologies to solve such problems, researchers have turned to biological systems for inspiration and a number of adaptive techniques for solving complex and dynamic problems have been developed. The following section will focus on some aspects of these paradigms.

1.27 Artificial intelligence paradigms

There is a tremendous interest world wide in the implementation of intelligent machines, which can mimic the behaviour of humans in a limited way. This process of creating intelligent machines is called artificial intelligence. The main paradigms for generating intelligence in machines include expert systems, artificial neural networks (ANNs), fuzzy systems, evolutionary computing, agent systems as well as the fusion of these paradigms. The main attributes of intelligence are learning or generalization, adaptation, fault tolerance and self-repair, and self-organization. The combined use of these complementary paradigms is also becoming popular and thus paving the way for the emergence of hybrid intelligent systems.

1.27.1 Knowledge-based systems (KBS)

The knowledge-based systems are software systems that can mimic the performance of a human expert by transferring his/her expertise to the computer in a specific domain. The major logical components of an expert system are a Knowledge Base, an Inference Engine, an Interface between the system and the external environment, an Explanation facility, and a Knowledge Acquisition facility.

A "knowledge engineer" gathers the expertise about a particular domain from one or more experts, and organizes that knowledge into the form required by the particular expert system tool that is to be used. (Barr & Fiegenbaum 1981) list seven different forms of knowledge representation: logic, procedure, semantic network, production systems (rules), direct (analogical), semantic primitives, and frames and scripts. A typical expert system uses a data structure as the basis of the particular representation technique it implements. Consequently, the knowledge engineer needs to know the general form in which the knowledge is to be represented and the particular form of that representation required by the expert system itself. The "engineered" knowledge is the Knowledge Base.

The Inference Engine is the "driver" program. It traverses the Knowledge Base, in response to the observations and answers provided to it from the external world, in order to identify one or more possible outcomes or conclusions. The data structure selected for the specific form of knowledge representation determines the nature of the program created as the Inference Engine.

The Interface between the expert system and the external world is generally provided by a user sitting at a keyboard and screen display. However, there is increasing use of input from other sources. Sensors and transducers of many different kinds are being used to monitor the environment; these are regularly interrogated, and their output used by the expert system. Output from other computer programs, including expert systems, is often used. Extensive use is being made of vast quantities of information stored in databases to which the expert system has access. The story is similar with respect to output, which may be directed to actuators, machine controllers, other computer programs, speech synthesizers, and external data stores.

The Explanation facility is often quite simple, merely being a listing of the inputs, the appropriate parts of the Knowledge Domain, and the outcome or conclusion. Very are explanation facilities provided to explain adequately the reasoning process involved. In some cases, ES components enable the knowledge engineer to provide special stores of text, which may be accessed by users who need further information, either about the questions being asked or about the conclusions that are provided. These suffer from the disadvantage that they must be fully created in advance by the knowledge engineer, who must anticipate the kinds of things about which explanations might be required.

In the case of the production (rule) based systems, the Knowledge Base consists of a number of rules written to an ASCII file. Therefore the Knowledge Acquisition facility required for these systems is often merely an external text editor. In systems based on other forms of representation, the Knowledge Acquisition facility will generally be an integral part of the expert system and can only be used with that system. The initial Knowledge Base is created using the appropriate facility, and modifications are made in a similar way. In very few systems does the Knowledge Acquisition facility possess any capability to "learn for itself"; the knowledge engineer is responsible for the maintenance of the Knowledge Base. The exceptions include those rule-based expert systems that have an "Induction" facility, often based on the ID3 algorithm developed by Quinlan 1983). A series of examples is provided to the system. Each example, which is generally the result of a specific observation, consists of a set of attribute values associated with a particular outcome. Provided that enough examples are provided, a decision tree can be "induced", and this can be converted into the rule format required by the expert system being used. Another approach is (RuleBook 1994) by Exsys, Inc. in which the knowledge engineer works with an expert to construct a decision tree graphically. The contents of the graph are then converted to rules.

The knowledge-based systems offer advantages over the conventional programs. Some of these advantages are cost reduction in achieving a complex task, reduced downtime, capturing scarce expertise, operation in hazardous environments, and, operation under incomplete and uncertain environment.

In spite of the limited success, difficulties associated with the knowledge-based systems include shortage of expertise and difficulty in extracting knowledge from experts. A main limitation of the expert system approach arises from our lack of understanding about cognitive processes and the way experts actually perform the tasks they do so well.

Knowledge-based systems have been applied successfully in a large number of applications. These include aerospace applications, design, diagnosis, interpretation of data, monitoring and prediction.

1.27.2 Artificial neural networks (ANNs)

A general introduction to the subject of Artificial Neural Networks is given in this chapter. The development of artificial neural networks has been marked by periods of considerable optimism and others of disillusionment. A realistic assessment of the potential of artificial neural networks is attempted and some of the unrealistic expectations which have grown around a new and developing subject are dispelled.

The study of artificial neural networks originally grew out of a desire to understand the function of the biological brain. This relationship between the biological neuron and the artificial neuron has been of great importance in past research but at the present time it does not appear to be a fruitful field. A basic description of the biological neural network is included so that the debt the artificial neural network owes to the biological neural network may be appreciated. The basic drive which has led to the development of ANN was a desire to understand the behaviour of the brain. The first attempts to model the behaviour of individual neurons in the 1940's were hardware based as these early studies predated the ready availability of the digital computer.

Most of problems at now able to be solved by means of artificial neural networks are generally also able to be solved by alternative methods. The question whether to use artificial neural networks to solve a particular problem is a matter of judgment on the part of the designer responsible for the project. The neural network would be a suitable candidate for use if significant advantages in important areas such as cost, speed of operation, reliability, ease of maintenance, ease of initial development, ease of deployment and modification can be shown to exist.

Artificial neural networks, usually called neural networks, are computers or software simulations containing elements that behave somewhat like the nerve cells in the brain. Artificial neural networks, which are electrical analogue of the biological neural networks, on the other hand can learn facts, represented by patterns, and determine the interrelationship among the patterns. The process of determining the interrelationships among patterns, in general, is informally called encoding. Some ANN parameters, usually the weights and the thresholds, are generally encoded to represent the interrelationship among the patterns. The process of determining the inter-related pattern(s) from one given pattern using the encoded network parameters is usually called recall. Depending on the type and characteristics of the encoding and recall process, the learning techniques are categorized into three major categories as supervised learning, unsupervised learning, and reinforcement learning.

Neuron models

The biological neuron is a complex cell that incorporates both biological and signal processing characteristics. The cell body is approximately spherical or roughly pyramidal in shape and it contains the cell nucleus and the mechanisms necessary for the survival of the cell. While every cell of a particular kind is an individual and has its own particular arrangement it is still generally similar in shape and arrangement to other cells of the same type. The dendrites form the physical means by which incoming information is received by the cell. The dendrites are arranged in a tree like structure, which has many branching processes. Neural Networks appear to have the property of interpolation but not that of extrapolation.

The name Artificial Neural Networks given to the study of these mathematical processes is in a sense unfortunate in that it creates a false impression, which leads to the formation of unwarranted expectations. Despite some efforts to change to a less spectacular name such as Connectionist Systems it seems that the title Artificial Neural Networks is destined to remain. At this time the performance of the best ANN is trivial when compared with even the simplest biological system (Allen 1994). The axon is the output channel of information from the neuron and extends away from the body of the cell. The axon divides into a series of processes, which in turn make connection with other neurons in the interconnected system. The axon is longer than the dendrite processes and in some instances may be a meter or so long.

Connection between the axons and the dendrites of individual neurons is the most frequent form of interconnection between neurons in the brain. While axon-dendrite connections are the most common form of connection, other forms such as dendrite-dendrite connections and axon-axon connections have also been observed by the researchers.

Neural network structure

The fundamental building block of the brain is the biological neuron. The shape of each neuron is distinct as are all the other component parts of each neuron. Each neuron is therefore unique as no two neurons are ever identical. It is estimated that the human brain contains somewhere in the order of 10^{11} neurons (Allen 1994).

The layered arrangement of highly interconnected neurons in most neural network models is largely based on the laminar structure of the biological cortex. At the simplest level, the cerebral cortex (or neocortex) consists of six basic layers or sheets of cells with functionally diverse and spatially specific interconnectivity between layers. In addition there are extensive lateral connections within layers.

In general, ANN structures may be described in terms of the interconnection method. For example, a multilayer perceptron uses connections between successive layers but not within layers, competitive learning models rely on competitive and co-operative interactions within layers (Kohonen 1988) and more complex paradigms such as Adaptive Resonance Theory (ART) (Carpenter and Grossberg 1987) employ both feedforward and feedback interconnections between layers and use interactions within layers as well. It is primarily the variety of possible interconnection structures that leads to the wide variety of properties and applications of ANNs.

Learning rules

A key capability of ANNs is their ability to incorporate knowledge in the weights through learning. A learning rule is a prescription to modify the strengths of synaptic weights between neurons based on external stimuli to the network and the resulting neural activity. There are two general categories of learning in neural networks, supervised and unsupervised learning. It is also possible to encode information into network weights using predetermined values (so called hard wiring), but this does not really involve network learning.

In supervised learning, both the input stimulus and the desired response to that stimulus are available and are used to formulate a cost (eg. error) measure. The derived learning rule in this case results from modifying the weights to minimize the cost in some way.

In unsupervised learning, the external stimulus is usually only available at the input layer and the weight modification mechanism incorporates predefined rules to drive it to a desired result. For example, in the ART paradigm, a mismatch parameter, the vigilance, is preset to allow the paradigm to decide either to incorporate information from reasonably closely matched new input into existing weights or to create a new category if the input is far removed from previously learned information. Thus the term "unsupervised" learning is slightly misleading since even in this mode of operation there must still be some external guidance as to the desired outcome.

Multilayer perceptron (MLP) neural network

The multiple layer perceptron (MLP) is a widely used neural network. It is formed by connecting more than two layers of neurons to each other. Figure 1.1 illustrates the architecture of a typical three layer neural network. This structure can be extended to the more general case of an L layered network of sigmoidal neurons where the layers are indexed by l; l = 0, 1, ... L. The outputs of the input layer are connected to the inputs of the hidden layer. The outputs of the hidden layer are connected to the output layer. This MLP can be trained to perform a given mapping. A number of training methods are used to perform this task. The gradient descent method of training involves minimizing the sum squared error (SSE) by feeding the error back to the input to readjust the weights.

A MLP has to be trained to perform a certain task. The purpose of this training is to obtain a minimized error in the outputs of the network. The so called gradient descent method involves minimizing the sum squared error (SSE) by feeding the error back to the input to readjust the weights. The weights have to be adapted in the way that the gradient of the error is negative:

Multiple layer perceptron (MLP) is a widely used feedforward neural network (Jain, 2001). It is formed by connecting more than two layers of neurons to each other. The outputs of the input layer are connected to the inputs of the hidden layer. The outputs of the hidden layer are connected to the output layer. This MLP can be trained to perform a given mapping. A number of training methods are used to perform this task. The gradient descent method of training involves minimizing the sum squared error (SSE) by feeding the error back to the input to readjust the weights.

A MLP has to be trained to perform a certain task. The purpose this training is to obtain a minimized error in the outputs of the network. The so called gradient descent method involves minimizing the sum squared error (SSE) by feeding the error back to the input to readjust the weights.

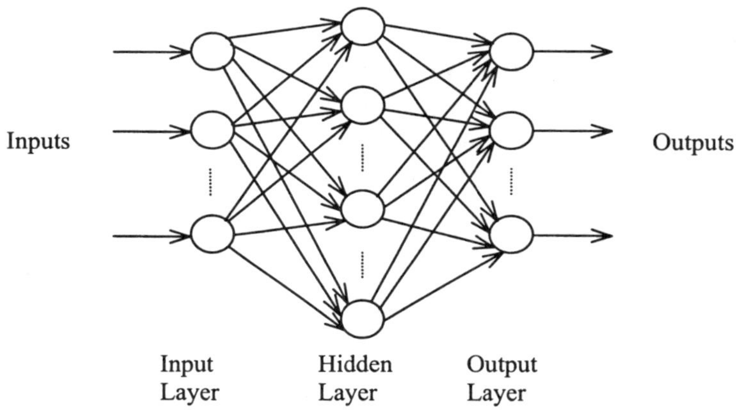

Figure 1.1. A typical neural network with one hidden layer and output layer

The weights have to be adapted in the way that the gradient of the error is negative. In a multi-layer perceptron, the error of a neuron propagates back from the output neurons to neurons in previous layers. That is why the gradient descent method is also referred to as the *backpropagation* method. Because the adaptation is made after all training vectors are presented, this kind of learning is called *global* or *batch* learning. It is also possible to adapt the weights after presentation of every single training vector. This is called *local* or *pattern* learning.

Neural networks operate in parallel and are modeled after the human brain. These networks are increasingly successfully used in various applications including areas as diverse as image processing, optimization, fault diagnosis, speech recognition and text recognition.

Neural networks have shown real promise in solving problems, but there is not yet a definitive theoretical basis for their design (and perhaps there never will be). The development process that has been used to date has been a genetic one, where the paradigms that have prospered are those that have been used by the most people, with the most success and on the most problems. Some of the more "unfit" paradigms are showing signs of withering on the vine. There is also evidence of evolution and crossover in paradigm designs as users experiment with various configurations and combinations of existing paradigms in an effort to devise new structures. If the engineering community is already using genetic techniques (even if not deliberately) to advance the field, my reasoning is that more formal GA techniques should be applied to see where the process can take us.

Neural networks have shown capability on smallish problems, but have not yet been able to solve the really complex problems that their biological counterparts can (eg vision). I see a need for automated design techniques that can help us to handle complexity. GAs possibly offer a method for doing that and at the least, I would hope to see us gain some insight into alternative approaches to NN design for complex problems that can come out of GA research. As a justification for this point of view I point to the inspiration that has been obtained from biological NN structures. Even though the hardware used by biological processors is vastly different from our computers, we have still obtained some fundamental insights from the way that biological systems solve problems. I hope that by evolving "silicon" solutions to problems we will find new signal processing insights.

We have some real problems which need to be solved right now and which cannot be adequately handled using conventional neural network design techniques. I wish to find out if any structures can be evolved that will solve them. These problems are specific to our area of work and have proved to be too complex for conventional NN techniques. However, similar problems are solved quite well by biological systems and we need tools that will allow the application of genetically evolved processes to the solution of these problems.

We have already done a considerable amount of work with neural networks and have significant expertise in MLP, PNN, Kohonen, ART and Neocognitron. We also have a number of moderately difficult real world test problems that we have already solved using ANNs. The approach we want to develop at this stage is to build up an understanding of GA approaches to network design using the standard problems and comparing the results with the known ANN solutions.

1.27.3 Evolutionary computing

Evolutionary computation is the name given to a collection of algorithms based on the evolution of a population toward a solution of a certain problem. These algorithms can be used successfully in many applications requiring the optimization of a certain multi-dimensional function. The population of possible solutions evolves from one generation to the next, ultimately arriving at a satisfactory solution to the problem. These algorithms differ in the way a new population is generated from the present one, and in the way the members are represented within the algorithm. Three types of evolutionary computing techniques have been widely reported recently. These are Genetic Algorithms (GAs), Genetic Programming (GP) and Evolutionary Algorithms (EAs). The EAs can be divided into Evolutionary Strategies (ES) and Evolutionary Programming (EP). All three of these algorithms are modeled in some way after the evolutionary processes occurring in nature.

Genetic Algorithms were envisaged by Holland (Holland 1975) as an algorithmic concept based on a Darwinian-type survival-of-the-fittest strategy with sexual reproduction, where stronger individuals in the population have a higher chance of creating an offspring. A genetic algorithm is implemented as a computerized search and optimization procedure that uses principles of natural genetics and natural selection. The basic approach is to model the possible solutions to the search problem as strings of ones and zeros. Various portions of these bit-strings represent parameters in the search problem. If a problem-solving mechanism can be represented in a reasonably compact form, then GA techniques can be applied using procedures to maintain a population of knowledge structure that represent candidate solutions, and then let that population evolve over time through competition (survival of the fittest and controlled variation). The GA will generally include the three fundamental genetic operations of selection, crossover and mutation. These operations are used to modify the chosen solutions and select the most appropriate offspring to pass on to succeeding generations. GAs consider many points in the search space simultaneously and have been found to provide a rapid convergence to a near optimum solution in many types of problems; in other words, they usually exhibit a reduced chance of converging to local minima. GAs show promise but suffer from the problem of excessive complexity if used on problems that are too large.

Generic algorithms are an iterative procedure that consists of a constant-sized population of individuals, each one represented by a finite linear string of symbols, known as the genome, encoding a possible solution in a given problem space. This space, referred to as the search space, comprises all possible solutions to the optimization problem at hand. Standard genetic algorithms are implemented where the initial population of individuals is generated at random. At every evolutionary step, also known as generation, the individuals in the current population are decoded and evaluated according to a fitness function set for a given problem. The expected number of times an individual is chosen is approximately proportional to its relative performance in the population. Crossover is performed between two selected individuals by exchanging part of their genomes to form new individuals. The mutation operator is introduced to prevent premature convergence.

1.27.4 Fuzzy logic

Fuzzy logic was first developed in the mid-1960s for representing uncertain and imprecise knowledge. It provides an approximate but effective means of describing the behavior of systems that are too complex, ill-defined, or not easily analyzed mathematically. Fuzzy variables are processed using a system called a fuzzy logic controller. It involves fuzzification, fuzzy inference, and defuzzification. The fuzzification process converts a crisp input value to a fuzzy value. The fuzzy inference is responsible for drawing conclusions from the knowledge base. The defuzzification process converts the fuzzy control actions into a crisp control action.

It is argued that the attempts to automate various types of activities from assembling hardware to medical diagnosis have been impeded by the gap between the way human beings reason and the way computers are programmed. Fuzzy logic uses graded statements rather than ones that are strictly true or false. It attempts to incorporate the "rule of thumb" approach generally used by human beings for decision making. Thus, fuzzy logic provides an approximate but effective way of describing the behavior of systems that are not easy to describe precisely. Fuzzy logic controllers, for example, are extensions of the common expert systems that use production rules like "if-then." With fuzzy controllers, however, linguistic variables like "tall" and "very tall" might be incorporated in a traditional expert system. The result is that fuzzy logic can be used in controllers that are capable of making intelligent control decisions in sometimes volatile and rapidly changing problem environments.

Fuzzy logic techniques have been successfully applied in a number of applications: computer vision, decision making, and system design including ANN training. The most extensive use of fuzzy logic is in the area of control, where examples include controllers for cement kilns, braking systems, elevators, washing machines, hot water heaters, air-conditioners, video cameras, rice cookers, and photocopiers.

1.27.5 Artificial agents

There is a tremendous interest in the design and applications of agents in virtually every area including avionics, business, engineering, health sciences, management and so on. There is no agreed one definition of an agent but we can define an agent as a computer program that autonomously acts on behalf of the use. In practice, varieties of artefacts such as synthetic lifelike conversational characters on the computer display, computer programs and broker information on the information network, or computing units that collectively exhibit an intelligent behaviour, are called agents (Jain, 2002). Intelligent agents are expected to mimic the behaviour of humans in a limited way. It is expected that the multi-agent system, will encapsulate the human's requirements and executes it with little guidance. This will require the learning and communication capabilities between various agents. Obviously, we need to design agents with the characteristics including independence, learning, cooperation, reasoning, intelligence and communication. It is important that the agent should be able to sense its environment and act autonomously upon it.

The learning and intelligence requires the development of artificial intelligence paradigms including expert systems, fuzzy systems, case-based reasoning and neural networks. Intelligent systems have a natural synergism that can be exploited to produce powerful components of multi-agent systems. A theme that can be found in these alternatives is the attempt to make up for deficiencies in the conventional approaches. In some cases, the goal is better, more efficient and effective multi-agent systems. Obviously, this requires adding features associated with human intelligence such as learning and the ability to interpolate from current knowledge. The appropriate use of intelligent technologies leads to useful systems with improved performance or other characteristics that cannot be achieved through other methods.

1.27.6 Summary and problems

The new problem solving tools of adaptive mechanisms provide an expanding opportunity for future systems to progressively assume more and more of the functions normally performed by human beings. The boundaries of problem solving complexity are continually expanding as researchers concentrate on biologically inspired techniques and the capability of computing hardware is also growing rapidly enough to underpin the advances in programming techniques. The new adaptive techniques signal a watershed in the progress of engineering, since practitioners have now begun to use computers as more than just tools to implement algorithms that have been developed by humans. We are now beginning to program computers in such a way that the software can work out by itself the best way to solve a given problem. This new approach brings with it a variety of new problems, not least of which is the question of how best to validate the performance of such software. For example, the behaviour of an ANN under all possible conditions can only really be evaluated by testing it under all possible conditions. This is rarely possible and so the network can never be said to be fully characterized. Under some circumstances this may be satisfactory, but there will be numerous applications where the certainty of an algorithmic solution will be required (the old question that is often asked to illustrate this point is "would you fly in an aircraft piloted by an artificial neural network?").

We should also expect that other members of the society will not share our enthusiasm for the new technology, since, not only does it signify a major change in how we use computers, but it also will have an impact on the work practices of others. As we become more competent in our use of biologically inspired programming techniques, we will find that the systems we develop are displacing people from their jobs. It is after all generally an economic imperative that is ultimately behind what we do and the outcome cannot be of benefit to everyone. We suggest that it is our responsibility to ensure that the results of our work with the new technology is widely publicized so that the rest of society may be prepared for its impact and only those applications may be selected which are found to be appropriate and wise.

Chapter 2

Physics of language

2.1 Communicative attitude, cognitive aptitude and textual actitude

There are many good terms suitable for defining most phenomena and one should carefully examine all of these before thinking of creating a new term yourself. This is rule number one as we have many words available already to define objects and concepts. In spite of this recommendation, it was decided to introduce some new terminology here and to draw a clear cut distinction between "attitude" and "aptitude" and "actitude".

Relying upon the basic assumption about the natural and consistent ability of individuals to trace back their communicative positioning attitudes and to recognize other individuals' communicative positioning attitudes, we can define cognitive aptitude as the cognitive ability that allows individuals to consistently recognize what is occuring in language as a reflection of what is possible owing to a commonly shared cognitive structure.

For example in some cultures and languages such as Japanese may a silence carry a whole message. This is of course only a message for those who know the set of possible meanings, which the act of silence can carry in Japan.

Cognitive structures, which is the way our mind is organized, allows different kinds of cognitive aptitudes. This allows different communicative attitudes to occur and to reflect upon communicative behaviour in general and upon natural language based performance more specifically as communicative patterns, which may be recognized as such, because they are cognitively grounded and therefore commonly shared by a certain community. As such they manifest themselves in texture, still at a very dynamic level and in text as a final state in statistical way, which can be traced back.

According to this view of language this newly and precisely defined terminology introduced, as well as the visual representation system, are specifically aimed toward creating a possibility for exploiting and exploding the visibility of information complexity entailed in any communicative intercourse and reflected in texts and documents too as a consequence. Some of those phenomena, which may be found in a text, may be only observed and carefully identified through an accurate process of self-reflection.

Transparency of such important processes and phenomena represents a significant condition for a consistent explanation of ambiguity up to misinterpretation, which is most likely to occur especially when communicative partners of different cultural background and languages communicate through the same language, as they may end up transferring their originating cognitive aptitudes and textual attitudes into a foreign language.

A visual apparatus as to convey communicative intentionality, as to complement text, is a powerful device for accurate understanding of communicative energies that are active in text, which may then produce consistent understanding. This is how the peculiar term "actitude" was used. This term applies to text to indicate communicative energies that actively produce consistent understanding.

Textual actitude is the communication and information energy potential. It may be left visible at the surface level, as it appears in a textual pentagram, where text is being complemented visually by its execution cues, thereby allowing for consistent intepretation to occur. Need for a change in interpretation may also be indicated and made explicit by using the visual apparatus dynamically, like suggesting how to pass from one style to another, by declaring the nature of the changes intended through the corresponding visual representation, either on a local or global level.

An example could be a certain joke or paradoxical expression which was spelled and then accidentally written down as a statement does become on offense or affront as taken out of its originating context.

Going back to a metaphor once again, this entire way of looking at language and coping with inherent complexity of any communicative intercourse, looks somehow like some kind of text X-raying process meant to have each detail surface and become visible.

But there is another, extremely powerful way to envision communication energy, similar to what happens in nuclear energy.

And this is precisely the possibility of manipulating previously processed paragraphs and sentences which we can also name as textual units, in order to first compress them, which is make them more compact and simplified as for their syntactic structure, as to then proceed toward another kind of reaction like acceleration and aggregation, which looks like fusion or fission processes after all as to keep the metaphor going.

2.2 Full exposure and obscuration in text processing

The illustration of a textual pentagram may be thought of as full exposure to textual communicative energy in the sense that it makes visible what is usually left invisible, which is the very purpose of the text. In the theory of text comprehension there are other ways to further exploit and enhance visibility of those qualitative aspects of communication, which are normally left invisible in text and may therefore remain obscure.

Full exposure to qualitative aspects of information may lead toward the need for clarification of each single sentence and paragraph within a certain text as to make sure that the intended communicative energy is understood as resulting into its actual textual forms.

For example an information specialist may very often have to confront the fact that a certain part of a document may diverge substantially from the rest part of the document, in ways which trigger the need for in depth search as for the intended meaning of that one section of the document.

Let's now consider the term "obscuration" in the way it is presented here. By obscuration we may mean the act of discovery that a certain part of a document is really inconsistent as for its nature and may have been subject to manipulation as for its originating context.

We may say that we have found a certain part of a document to be obscure as a result of an obscuration process performed on it. By obscuration we may also mean the process of intentionally eliminating contextual information which could be of relevance as to understand and capture the real meaning of a certain sentence or paragraph such that a part of a document up to the whole document may become obscure as a consequence or totally lacking as for interpretive light or illumination. This is a deliberate evasion of clarity in two different ways:

A text or textual segment is obscurated if

a. some parts of it are lacking qualitative information cues for consistent interpretation,
b. some changes in text interpretive cues do not show traces of the originally existing interpretive cues. Thus the textual structure under examination may be considered to be missing the interpretive light and illumination.

These kinds of phenomena are intrinsically extremely dangerous because they may cause partial or even total loss of the originating information sources, this way making verification of reliability of information impossible in the first place.

What the system here illustrated does is to keep all the information energies transformation and execution cues visible and available all the time.

Awareness as for possible risks, which may occur if obscuration takes place, both intentionally and unadvertently must be provided as to raise sense of responsibility and commitment on the part of the information scientists and information specialists who do play a most delicate and most important role in the world of information overload.

2.3 New evidence, new theories and new terminology

In physics it is extremely important to take into account the theory behind each experiment and its model. Many important observations made by physicians in the past are lost because we do not have information about their framework of interpretation. In physics, any good experiment is based on the tools the observer has used and on conceptual structure upon which the observer has reframed experiences and experiments. It is also very important to distinguish between "prejudiced observation of facts" and "high sensitivity toward phenomena observed "; the first attitude being extremely damaging for any kind of scientific analysis, the second being the correct attitude toward analyzing a certain experience so as to achieve a detailed knowledge of a certain fact or phenomenon.

A consistent perception of a certain fact or phenomenon in science should lead toward accurate analysis and possible reproduction of the same or of an equivalent. High perception will only be possible if it is grounded upon a clear-cut, coherent and concise symbolic representation system of the phenomenon, which is subject to scrutiny.

What has been here presented and defined in terms of "textual perception" is exactly the kind of attitude any information scientist and specialist should be able to evolve first and then to adopt all the time in order to observe, identify, select and interpret information coming today from all kinds of contexts and set ups.

"Textual perception" may be significantly enhanced by a consistent system for visualization and annotation.

There is a clear distinction as for identifying common sense laws of meaning, physical laws of syntactic structure and communicative energies of pragmatic nature, which are always active when we talk and write and obviously need to be conserved and preserved, as to be recognized by our listeners and readers.

Some observation of facts may be enough to establish "true or false" conditions as for those statements, which involve common sense based on agreed and consensually shared experience, as someone who writes about a certain political topic in France may assume that a French readership will have some background knowledge as to be able to follow.

As for more impalpable aspects of information such as intentionality may be, we need to really to address the problem differently. Some kind of physical laws of language may not based merely on experience but on experiments as well, which means that different conceptual tools and observation criteria will have to be adopted at different times.

Specific nature of this kind of tools is therefore intrinsically different and the "true or false" condition may not always be considered as a valid classification criterion as to decide about the adequacy of a certain model for interpretation of a certain document.

What really matters is the power of the verification mechanism, which is based both upon experience and observation.

And the degree of approximation may vary considerably according to the power of those observational tools, which are being used.

Research on common sense and knowledge representation in language is based on a stable "platform" that rests on the fact that abstract terms, properties and attributes are linked to facts and experiences.

This is why we see a relatively fast development of highly productive knowledge representation models and theories in the field of artificial intelligence, based on strong and powerful concepts such as frame (Minsky 1975), script (Schank 1977), plan (Wilensky 1983) and plot unit (Lehnert 1981).

A law in common sense is a general assertion which may be either true or false and which can be tested and confronted with so called reality quite directly and unambiguously. As a result of immediate testing, it may be declared to be either correct or incorrect or incomplete, and if it is accepted, it will be added to a whole other set of assertions, which may well create a repository of solid knowledge.

Research aimed toward exploring energies and forces active in language, both at a syntactic level and at a pragmatic level, is based on symbolic representation, which may be more or less adequate and complex and is necessarily continuously evolving, as for the state of evolution of the theory it is based upon. Any symbolic representation is in fact strictly connected to a certain theory; this is why it is so important to consider each term and element in the very context of the specific theory been adopted. Otherwise much confusion and ambiguity may be generated as a result of meaning attributed to a certain term that is used differently according to a different perspective in a different domain.

2.4 Old evidence, new evidence

It is very important for any scientist, and particularly for an information scientist and information specialist, to be aware of the terminological consequences derived from newly acquired evidence about certain facts and of the possible ambiguity that ordinary terminology may generate if the basic assumptions have been changed without previous advice.

If one has to classify the new framework as illustrated throughout this chapters as belonging to pragmatics tout court, the only aspect that would result into a correct interpretation would be the literal meaning of "pragma", meaning "activity" and "action"; the theory being based on the assumption that text is the basic communicative active entity. But Pragmatics as a specific branch of Linguistics today is mostly related to descriptive analysis of conversational interactions, discourse phenomena, specialized kinds of discourse and typologies of discourse and speech acts, hence it would be difficult to define this approach merely in such terms.

Further, there is an animated and interesting discussion continuing and ongoing in the field of Linguistics about what a "text" really is and means as opposed to "discourse", resulting in a wide variety of opinions and options and categorizations and subcategorizations. The present view about language and information radically differs from others because of the introduction of a quite complex view, which may productively defined as some kind of "quantum mentality" as for what language analysis is. Text is here in fact considered to be the basic complete and active communicative entity. A quantum theory of language and information is also meant to reflect some substantial changes of perspective, as it is aimed toward identifying first and then link up together the qualitative and the quantitative aspects of information materializing into texts.

The chomskian dichotomy between "deep structure" and "surface structure" may still be considered to be a valid one because it covers those substantial differences between the "potential" and the "actual", though obviously would Chomsky apply such dichotomy to sentence, which means the syntactic structure.

As a consequence of such productive distinction extended toward the pragmatic dimension, a further dichotomy between "texture" and "text" has been here introduced and considered to be a productive one.

The verb "cognize" was introduced by Chomsky to mean "knowing the grammar of a language", the verb "know" being too general and therefore ambiguous.

Along the same lines, the verb "recognize"is introduced as to mean precisely "know the positioning of a certain text based on a commonly shared cognitive program or structure".

As we know well specific terminology in a scientific domain may either be reframed or extended or even fully replaced, when a certain term has such a long tradition that adding one more meaning to it may result overwhelming.

Adding one more definition would certainly slow down the process of correct interpretation and use, resulting into an ineffective way of identifying and expressing newly acquired evidence. Nevertheless, leaping some connection with a previously existing framework for interpretation and scientific model may be productive after all especially if this model is well know and well established.

Classical physics and quantum physics have shown so many commonalities that some apparently contradictory concepts have been found to be compatible after all, as in the case of corpuscular and wavelike theories of matter.

Thus, it is not suggested that this framework is contradicting up to disconfirming the classical statements of linguistics or knowledge representation research in any way, as it is rather a compatible one meant to support tools for most accurate interpretation of communicative events.

2.5 Abstract theories and mechanical models

If we analyze thoroughly what has been going on in physics for centuries, we can see an alternation of abstract theories and mechanical models, developed to explain observable or unobservable facts and events. If the main issue in the science of language today is not so much categorizing facts and events as this has been done so extensively before, rather creating a consistent representation of what is behind language facts and events, there should be no contradiction at all in alternating and harmoniously combining abstract theories and mechanical models as to illustrate specific facts and processes.

No doubt that just as for any science we may not think of an exclusive and fully comprehensive theory of language, but we have to consider a set of different theories acting at different levels, which may be productively applied as to solve different kinds of problems.

Generalizing some observations correctly and identifying common rules will result into the consistent set up of abstract theories bound by a need for economy, whereas mechanical models will rather be aimed toward capturing a certain fact found to be quite complex, by reproducing some of its aspects or by modeling it as for analogical reasoning. Being able to find or produce most consistent models for representation turns out to be a very important issue.

Within the current framework here presented, visual representation is specifically intended to fulfill explanatory needs and each visual scheme may easily be turned into a set of operational devices for practically solving qualitatively different problems. Since language facts and communication processes, which need to be analyzed, may be both simple and complex, those visual representation systems designed do reflect such structural difference and they are therefore qualitatively different. For this very reason they differ enough not to be repetitive, redundant or overlap.

Textual canvases and textual machines are aimed toward representing complex, dynamic and therefore "scruffy" processes in the preliminary phase of text planning and design.

Textual signs and textual symbols are designed to represent the final result of such processes in a stable, detailed and "neat" way.

Such diversity of representation systems does actually provide a full guarantee as for accurate and comprehensive description of inherently different communication phenomena and processes, which of course need to be viewed as separate ones and also be distinguished from one another.

One more aspect needs to be pointed out: simple facts may result into more complex combinations and may need to be observed in order to be defined and explained when they are active as a combination. Facts, which look static at last may represent the end result of a dynamic set of processes which have previously occurred.

Such dishomogeneity is not to be considered as an inconsistency; it is rather a proof that correct definitions of dishomogeneous facts and processes are occurring. The term dishomogeneous is better replaced by a term such as "complex" is; not to suggest any negative connotation. If simple facts or processes may be isolated and identified as such, then a consistent representation may be found; if they tend to operate as a combination, a consistent representation should take into account the whole system of facts and processes in a dynamic way and a consistent and qualitatively different kind of representation need therefore to be provided and made available. Mechanical models are particularly suited in this case, because they may animate highly complex processes involving different energies and forces, which are active simultaneously, by still keeping the right progression. This is the reason of designing textual machines. If there is some overlap in those mechanical models, and just a very limited level of redundancy this should not cause concern; some of those processes, which are taken into account, may in fact be active in more than one of the described phenomena at the same time, and therefore may need to be represented in a variety of ways. What will therefore follow next may be considered to be a "realistic" representation, meant to support explanation of complex language phenomena; under the assumption that forces and energies of different kinds are always active in language at different levels. As for a quite effective metaphor coming from physics at this time, we may add that "language matter" may therefore change, altering properties at a structural level, with real changes occurring when exposed to such forces and energies – like iron undergoing magnetism. Forces and energies represent qualitative aspect of communication, and deeply and intrinsically affect language structure. According to the present framework, the need was always felt to talk about text in terms of an entity subject to differing energy states, which has indeed proved to be very productive in order to be able to discriminate consistently qualitatively different phenomena and processes undergoing observed.

Creating qualitatively different models to justify and explain qualitatively different communicative facts and communicative actions guarantees that qualitatively different problems will not be mixed up and handled arbitrarily.

2.6 Language, information and communication

All the way through this research path and work, we have referred back to etymology, which is the science of meaning and meaningful relationships evolving over time as through it, it is often possible to find direct evidence to support very strong scientific claims.

Information – from Latin "informo" – means literally to give shape, shape up. Information is the process through which knowledge, coming from different experiences, is organized in different forms and formats. As a consequence of such processes, information may be referred to as a certain kind of "matter". Natural language is one of the three major states or phases of that matter, which is information, the others being visual languages and gestural languages. Each state of information has properties that distinguish it from the others and is therefore independent of the particular "substance" being studied. By "substance" we mean specific linguistic entities as they become visible and therefore are subject to syntactic analysis in a particular language.

Semiotics has been largely concerned with the study and description of the different properties that characterize each of the major states or phases, which are currently named as "codes" (Eco 1975; Sebeok 1960). Semiotics as such has never been concerned with processes such as transformations occurring in ways, which are not directly visible, as it is largely concerned with interpretation of observable phenomena and consistent description of facts. A semiotic approach to communication is of course of great significance as to understand the complexity of elements involved into any communicative endeavour. What is really added here on top is a new way of looking at information as in the form of information units of different sizes and qualities.

A visual information unit will differ from a gestural information unit or a natural language information unit, just like a gaseous state differs from a liquid state and a solid state, yet all of them are "states of matter". Information may consist of simple and complex units which may be interpreted through a highly articulated process, involving a whole set of cognitive processes. The different states of information have properties which make them look different at some times, so that not even the same terminology can be used, because not even a common set of problems is being addressed.

Just to provide an example, it would not really make much sense to address the problem of summarizing a gestural sequence or a few icons as we would on the other hand certainly productively think of the possibility of summarizing a novel or an explanatory text.

We cannot think of accessing just one common system of interpretation since we would be referring to qualitatively different processes and entities. Drawing distinctions among different states of the same matter by applying radically different but still compatible systems for both representation and interpretation is a widely accepted, highly productive practice in physics; which is meant to avoid confusion and provide clarity. And of course the same phenomenon may be described and defined by two apparently different models that are compatible and, at the end, productive, as in the case of corpuscular and wavelike theories of matter.

Along those very same lines we may indeed think of many and different kinds of representation systems that can be applied to a very complex linguistic entity such as "text" is infact.

Qualitatively different kinds of interpretation and understanding are in fact possible, and the existence of such processes is supported by evidence. Interpreting and understanding a natural language sequence is substantially different from interpreting and understanding a visual language sequence or a gestural language sequence.

As to discriminate among such aspects of interpretation and understanding, we may postulate the existence of linear-sequential interpretation and understanding processes and global-vibrational interpretation and understanding processes. Such differences will become clear in the course of the exposition.

There are cases where information particles that have been identified in each of the previously mentioned information states may be considered to be relatively independent of the influence of nearby particles, whereas in other cases they may be found to be subordinated to forces exerted on them by neighbours. Forces and energies of different kinds apply to information sequences in many and different ways and they produce qualitatively different communication effects.

It is difficult to establish a clear distinction between information and communication, just as it is hard to draw a line between the particle-like and the wave-like nature of natural language. They are certainly very much interconnected phenomena: as such they manifest themselves in different but consistent ways in the three major states or phases that were first introduced, that is in natural language, visual language and gestural language.

Both concepts of information and communication (Latin "informo", shape up and "communis facio", put something in common) imply a dynamic process taking place in various steps and through different kinds of operations. In some ways we may add that communication forces and energies act upon information sequences in such a way as to turn them into "inform-active" units of discourse, where the information potential is actualized. The new concept of "inform-activity" will be further explored later in this book as for its various manifestations.

2.7 Communicative forces and energies active upon language

When certain kinds of forces and energies become active in natural language, not only do they affect and determine the communicative positioning of a certain text by reflecting upon stylistic choices pervasively but they also have an impact on the syntactic structure.

This certainly needs further direct exploration as to be practically used to solve some of the as yet unsolved problems. Text segments and text units of natural language in a certain communicative space or context may be considered to collide with one another and rebound to produce different kinds of interpretation and understanding.

A very important property of the natural language state of information is that it is able to represent any kind of context whatever communicative space it is put in, where communicative space means the context. Other important properties of the natural language state of information are the following:

(1) text sequences and full texts may be compressed at a local or global level when synthesis is applied;
(2) text sequences and full texts may be expanded when analysis is applied;
(3) text sequences and texts may be condensed up to a concept or a set of concepts, when abstraction applies.

If these changes take place, they affect the nature of the original text sequence or text in a way that will reflect on the interpretation and understanding of the text sequence or text. It is called "pressure" the way the communicative positioning of a text sequence or text reflects on the information structure of that text sequence or text, which is called "temperature". Pressure is therefore applied to the specific communicative component of text, whereas temperature is applied to the specific informative component.

A further statement can be made, which may be conceived as a kind of physical law of text temperature and pressure balancing, where any change of temperature leads to pressure changes, and vice versa. This is how the quantitative aspect of text, which is the information structure, and the qualitative aspect, which is the communication energy, may harmoniously be linked together. Changes in temperature and pressure occur all the time and they may now not only be noticed but observed, measured and analyzed very accurately within the new framework.

2.8 A macroscopic view and a microscopic view of language

In order to be able to observe and explain qualitatively different phenomena and facts about language, both a macroscopic view and a microscopic view are needed. A macroscopic view is specifically represented by the overall communicative positioning framework, which is certainly connected but also independent on an atomic theory of language like a syntactic one.

By "connected" we mean to indicate very important relationships that exist between positioning and its reflection upon the syntactic structures; by "independent" we mean that communicative positioning may be analyzed and explained by referring to internal and specific criteria. A description of textuality in terms of measurable properties may be defined as the "phenomenological thermodynamics of text" and "text transport theory".

A phenomenological thermodynamics of text is largely concerned with information distribution and information structure of each text segment, in other words with text temperature. A text transport theory is largely concerned with the reflection of the communicative positioning of each text segment, in other words with text pressure.

Both phenomenological thermodynamics of text and text transport theory nicely fit into the new framework presented here. More specifically, the phenomenological thermodynamics is represented with canvases and tapestry and fresco text representation systems, while the text transport theory is represented with textual machines. Textual canvases and tapestry and fresco on the one hand and textual machines on the other share the same property of representing macrophenomena, which are complex and "scruffy", and also represent processes evolving in different stages. They represent some basic actions taking place before the text is even shaped; whereas an actualized text keeps traces of such fundamental processes or movements which occur through a different kind of text representation system, which is a more stabilized one and refers back to textual signs and textual symbols.

Both kinds of processes have a real impact on the syntactic structure in a very positive sense: some phenomena that are hard to handle syntactically because they refer to something external – which Chomsky has recognized many times as part of "discourse" or referred to as pragmatics -, may be easily handled within a text phenomenological thermodynamics or a text transport theory so that when it comes to their syntactic manifestation they have already been explained. Some difficulties arising within a purely syntactic framework are due to the fact that some issues would be more easily solved within a communicative positioning framework instead.

A microscopic view deals with the description of each sentence phenomena, of specific language facts which are to be found on a syntactic level, and with explanation of rules regulating such phenomena.

A full description of a system of interacting rules is also required in order to explain the behaviour of a native speaker. The most powerful model of this kind now existing is Chomsky's framework, which only applies to syntax though. It would be useful to define his theory in terms of a kinetic theory of syntax, because Chomsky has identified movement and changes existing in the syntactic structure. Both a macroscopic view and a microscopic view are needed to explain highly complex phenomena, which manifest themselves in natural language.

2.9 Cognitive structures, knowledge structures, syntactic structures

The reason for postulating the existence of cognitive structures is that cognitive processes like focalization, selection and so on, create specific effects on natural language that are clearly different from those we can find active in visual or gestural languages. By this, if one says something such as "focus on that part of the image", it really means some cognitive operation, which is substantially the same but has different implementation modes according to the state or phase of information to which it applies.

Cognitive structures may therefore be defined as dynamic structures working and based upon cognitive processes manifesting themselves via natural language. Going back to the previous example, focusing is different in nature when it manifests itself in natural language or in vision: iconic focusing differs from textual topic focusing.

In order to differentiate and still retain the same terminology to define each of those cognitive processes, it was found necessary to refer back to the existence of cognitive structures, which are the output result of cognitive processes having been applied to the natural language state of information. It would be useful to abbreviate the term "cognitive structures" which is by definition plural because of the many different cognitive processes that are involved into the term "co-structures".

"Co-structures" are intrinsically dynamic; it is important to keep the dynamic aspect of representation active all the time in order to explain and justify complex phenomena, which are part of a more general linguistic theory. Co-structures become the dynamic grounding upon which language facts can be consistently built. They neither interfere nor diminish the role syntactic structures play; they rather reinforce and support them.

It would be useful to add that they make possible the actual functioning of knowledge structures such as the frame (Minsky 1975), the script (Schank 1977), the plan (Wilensky 1983), and the plot unit (Lehnert 1981). Knowledge representation structures such as the above need in any case to be activated at some point on the basis of a very complex organization, which needs to be a dynamic one.

Co-structures may allow both the syntactic representation component and the knowledge representation component to be harmoniously and synchronously active, with mutual coordination and no interference.

Only through harmonious merging of different components may we expect to handle such complex problems as language facts can provide us with. A set of physical examples of undefined and invisible problems such as those raised by language have been illustrated as part of a framework and based upon qualitatively distinct visual representation systems such as textual canvases and textual machines are. There is a need to create a physical example that may work like a model of the above mechanism based on different structures and which may allow a full interpretation and understanding of natural language texts. As to explain this concept, let us introduce the charging machinery for handling natural language. Let us now have a look at Fig. 2.1.

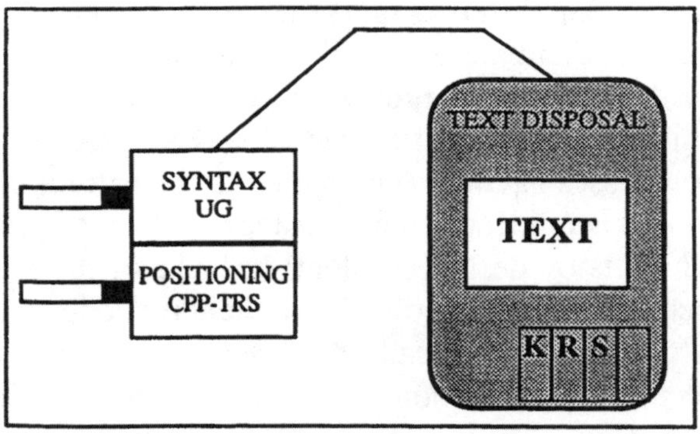

Figure 2.1. Charging machinery for handling information in natural language

The intention is to represent visually with the charging machinery for handling information in a natural language state is the built-in syntactic component, represented by the universal grammar (UG), and the positioning program (CPP-TRS), which are plugged directly into the mind; they are both linked to an external disposal which is made of the knowledge representation systems (KRS) that are acquired through experience and therefore not to be found in the mind originally.

When a text comes in, it will be charged on the text disposal. This will become active as UG and CPP-TRS fill with the energy that allows the KRS component to become active. The KRS component may be partially active even if UG and CPP-TRS are not plugged in but in order to get the full charging to reach interpretation and understanding of the text at some point, UG and CPP-TRS will have to be plugged in. This is all due to the diversified activation of syntactic structures, co-structures and knowledge structures, based upon different parameters.

In language understanding we can also differentiate among different kinds of expectations, which are respectively syntactic expectations, positioning expectations and knowledge representation expectations. Syntactic expectations may be handled within a kind of case grammar (Fillmore 1977), which needs still some further extension in order to cope with a wider variety of syntactical facts. Syntactic expectations may well be handled accurately and productively within a Chomskian framework.

Positioning expectations need to be accurately described. Some kind of positioning expectations may have to be found in the relationship, which exists between and among textual functional indicators on a global level. In other words, a textual sign may create the expectation for another one to come either before or afterwards. An example of expectation driven syntax for textual functional indicators is represented by the sequence inscribed arcs-opened text space. For such sequences we may think in terms of an "If-Then rule", whereas other sequences are weaker and certainly not deterministic.

Links between and among text styles indicators are almost non-existent and extremely weak. Links between text styles indicators and text turn-taking indicators are again extremely weak and almost non-existent. Links between textual functional indicators and both text-styles and text turn-taking indicators offer some rather weak predictions, but only tentative ones and certainly not deterministic ones.

Knowledge representation expectations have been widely explored within the artificial intelligence tradition of knowledge representation and of common sense; we certainly know that the strongest expectations are context driven.

2.10 Communicative patterns recognition

Before discussing the specific issue of linking communicative patterns recognition to co-structures, let us first go back to the question of familiarity versus unfamiliarity. Familiarity with knowledge structures, which determine the contextual meaning of each sequence, may certainly help identify the most consistent positioning, which may be derived out of an appropriate communicative patterns recognition process. Correct context assignment to each sequence is crucial. Let us now give an example, which comes from personal experience.

Graziella was at a conference in Italy, speaking to an Italian colleague who was also giving a talk. During a coffee break, Graziella wanted to tell him that she had submitted a proposal for a scientific paper that she had first published in English, to be presented in Italian also. Here is the conversation:

Author: I have just submitted a proposal for the translation of my paper, which is also somehow translated.

Colleague: Oh good. It is our old back-and forth between languages.

As to make out of this short conversation some sense, a very specific contextual information is needed. By attaching a correct context to each sequence, the most consistent positioning may also be found. In this case there is a context both shared: Italian scientists having to cope with the fact that they may have to work in two languages, as to be understood in Italy and to obtain an international audience. If we were to define the communicative positioning out of the sequence presented, the attribution would be that of "analogous story" as the two communicative partners share the same experience.

Conversational implications are of help but there is something more needed as to really achieve a deep and full understanding. In other words, pragmatics is merely acting in a classification mode and very much at a level of description as it deals with different contexts, which are categorized.

According to this new perspective, the so-called positioning represents a kind of pragmatic grammar meant to define different communicative actions, reflecting upon different languages, in different ways and which may still be recognized by the native speakers throughout communicative patterns recognition.

2.11 An explanatory theory of pragmatics

Extensive research going on today in pragmatics is really about context driven communicative patterns identification, observation and description. This is important and valuable work which corresponds to linguistic categorization search, which was previously done only at a descriptive level before the Chomskian revolution took place. Much observational data are important on a descriptive level, and up to now pragmatics has been largely concerned with descriptive matters of course.

In order to create an explanatory theory, however, what is needed is a highly complex and highly articulated system, which may have well defined parameters and points of reference. This is what has happened in the realm of syntactic analysis within Chomsky's framework. As a consequence of an increasing real complexity, which was progressively surfacing, can see how the same framework has evolved in time, sometime almost denying the basic assumptions upon which it was first based. The need to define a really consistent universal grammar has never been fully disproved by difficulties.

The current approach here illustrated intends to provide an explanatory theory of pragmatics in two different ways. First, a theory of text comprehension is an explanatory theory of language seen as an information phase, which may represent and explain naturally occurring communicative facts showing in texts.

The theory of text comprehension here presented deals with actual texts and actual performance. Second, a theory of text compression, which may be viewed as a consistent output of the theory of text comprehension, may become so powerful as to generate and transform naturally organized texts into compressed texts, which have undergone condensation processes of various kinds.

In some ways the currently defined approach to language activities may in fact support interpretation of naturally generated texts based on a set of interpretative devices, which act upon different levels and represent qualitatively different processes and phenomena.

And those same interpretative devices defined and visually represented are consistent so that they may be used in combination and also be used dynamically to indicate transformations occurring largely as reformulation up to full condensation of texts. As such they both indicate existing processes and may also determine and activate processes.

As a consequence of high consistency of representation, the actual model is powerful enough to generate its own texts as for compression. This is where the model presented may become a tremendously powerful mechanism for autonomous processes of transformation of textual sequences keeping communicative energy and information force active and ready to be modified through reactions, which may occur over time. Analogies with nuclear processes here become a very powerful descriptive device.

2.12 Deep knowledge of the material and mastering of different techniques

Professor Tonfoni has always been intrigued by the basic principles that determine the intimate structure of different substances and which change her intuitions into deep knowledge of causes enabling her to generate theories supported by operatively valid models. In order to do this, the "material" needs to be split and cut in order to be penetrated.

Lucretius in his book on "De Rerum Natura" provides a good example of visionary intuitions based on the construction of a metaphor which is then turned into a whole set of scientific descriptions and explanations of physical phenomena. Someone may wonder why a physical theory of language is to be linked to a set of interpretative techniques first presented through metaphors derived out of painting, architecture and music. It is a good idea to explain those links by referring to what happens in painting when the painter starts thinking about the chemical properties of his paints. It is one thing to paint with oil, another to paint with gouache. Each paint has properties that deeply affect the product that will result from using one instead of the other. Differing molecular structures in painting affect light reflections as well as changes that occur in the chemical nature over time. Analyzing the different properties of the various information states, specifically concentrating on natural language both at a macro-level and at a micro-level of observation is most important.

As for architecture, let us focus on the need architects have to know about the intrinsic properties of the material to be used, both physically and chemically. Nothing may be planned or even conceived if the nature of the material has not been taken into account and thoroughly explored first.

As for music, nothing is arbitrary: everything may be described, defined, conveyed and executed according to a very specific plan based upon a highly technical notation system. Switching harmoniously from one paradigm and frame of reference to another one may allow a scientific approach to be both rigorous and also independent.

The greatest Italian artists in the fifteenth century were all very much aware and knowledgeable as for the material they were using and familiar with a wide set of substances, arts, techniques and different ways of doing things. Further, they were scientists as well because they knew theories behind and beyond what they were doing. We should never forget Leonardo Da Vinci's motto "Studia prima la scienza e poi seguita la pratica nata da essa Scienza" (first study science and then go for practice and tools derived from it).

2.13 Communicative energy and text perception

Just as light may be considered to be "responsible" for sight, communication energy may be considered "responsible" for text perception, which means understanding of text. Text is the actualization of that specific state of information that is natural language.

It is interesting to analyze the interaction between communicative energy and information as for each of the three different states in which information may appear: natural language, visual language and gestural language. Let us now concentrate on interaction between communicative energy and natural language, which is actualized in texts and text sequences respectively.

Just as physicists talk about light propagation in terms of both particles and waves, scientists of language may consider wave-like phenomena of communicative energy propagation in text to be completely compatible with particle-like syntactic phenomena within the newly established physical theory of language.

Communicative energy is a form of energy conveyed through textual space at a certain velocity, without necessarily transporting the matter, which is the actual text or parts of it; traces are in any case left as to help one to recognize the nature of the communicative energy at each given time. Just as with light, a text or a text segment may absorb, reflect and refract communicative energy in a certain way before it may reach the text observer. The human textual perception system is able to extract a part of the information, which comes with the text or text segment.

Through the use of textual instruments such as those provided within this framework – textual machines, textual canvases, textual signs, textual symbols and textual objects – readers, who may be defined as able to raise their own textual perception significantly.

Let us define those tools as spectroscopic instruments for communicative energy and information detection, just as telescopes and microscopes may extract extremely important information from the environment, which the human eye alone could not possibly obtain.

When communicative energy is incident, in a perpendicular direction, on a text or sequence (i.e. that it shows itself explicitly with the corresponding definition), it may be transmitted directly. For example, this is what happens when showing a comment on a certain fact:

This is a comment on the actual political situation in Italy.

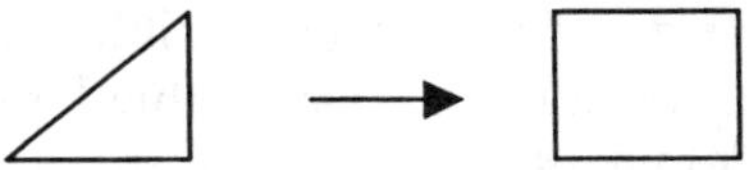

But this is rarely the case. Most of the time communicative energy reflects on a text or sequence and it is far easier to recognize communicative energy after a certain interpretative process has occurred, as in the following example:

This is an opinion about a political situation in Italy today.

With a particle-like theory of syntax, there would be serious problems in justifying communicative energy taking different forms in text. If communicative energy consists of a stream of particles, like lexical items or sentences may be thought of and because like cause should produce like effects, it cannot be explained why some particles are communicatively charged and others are not. With a wave-like theory of text it is easy to explain that the velocity of communicative energy transmission may vary because it is the result of wavelength frequency, which may vary as well.

This last point helps us understand the relevance of a particle-wave-like theory of text in the theory of text compression, which is based on qualitative reasoning about communication and information.

2.14 Mechanical effects of communicative energy

The relevance of qualitative phenomena and qualitative reasoning about text, which represents the entity to be analyzed in the information state of natural language, has been widely confirmed.

However, relevant quantitative changes and phenomena also take place in communicative energy transmission and should then be taken into account because they are extremely relevant. Such phenomena may be defined as mechanical effects of communicative energy. When communicative energy is either emitted or absorbed by a text or sequence in a text, three mechanical effects are likely to take place:

1. an exchange of information density in sentences or paragraphs;
2. an exchange of sentences and paragraphs linear distribution;
3. an exchange in the syntactic structure of sentences.

Those quantitative changes that may result in reductions, expansions or reformulations of the actual text may manifest themselves in the following ways:

1. topic-comment shifts;
2. subordination and coordination shifts;
3. ellipsis or elliptical phenomena.

These are quite radical quantitative phenomena, which are generated as a consequence of the qualitative phenomena, that were explored earlier. All of these effects may be explained partially within a classical particle theory of syntax and globally within a quantum theory of text. As a matter of fact, it is by referring to the particle-wave-like theory of text that those phenomena can be explored in relation to exchanges between communicative energy and both texts and text segments at a macro-level or as interactions between and among sentences at a micro-level.

An electromagnetic theory of language communication may well define communicative energy by explicitly recognizing and stating that each detector of communicative energy, such as textual machines, textual canvases, textual objects, textual signs and textual symbols are, may also be thought as being a communicative transducer and converter.

As for a quantum theory of language we may state that absorption of communicative energy is a continuum process.

Communicative energy may be exchanged in infinitesimal amounts between sentences and both coordinate and subordinate clauses by keeping information equilibrium, where equilibrium is a dynamic concept.

A sentence may exist only in one out of a set of discrete stationary states of information. Information has some internal energy of its own before it is charged by communicative energy as well; information energy is substantially different from communicative energy, so that it will be good to define it as informative force. Just as in physics, informative forces are able to determine acceleration, as we will see as far those text-acceleration processes, which take place as to determine text compression.

Chomsky (1980) stated that "linguistic knowledge, of course, extends beyond the level of the sentence. We know how to construct discourses of various sorts, and there are no doubt principles governing discourse structure".

Even though he has only focused on sentence and syntax, Chomsky has indeed recognized the need for explaining principles governing a qualitatively and not merely a quantitatively different linguistic entity such as discourse is in fact The need to further explore the qualitatively different nature of the components led me to the formulation of a theory of information activities as well as of a theory of communicative energy. It is possible in physics and in natural philosophy, as Faraday defines it, that various observable forces are intimately interconnected, for example electric current produces magnetic force and vice versa. Similarly in the physics of language, communicative energies affect information forces and vice versa.

As previously illustrated, natural language is one of the three major states or phases of information, the others being visual and gestural languages. Each state of information has properties that distinguish it from the others, and are therefore independent of the particular sequence or text analyzed. Languages consist of sentences, more or less complex, involving a lower or higher set of embedded clauses. Full understanding occurs through a textual space and is dynamic in its nature.

On a purely syntactic level of analysis, each sentence may be analyzed and parsed independently on the influence of nearby sentences; but at a textual level each sentence is subordinated to the information forces exerted by the wider textual space as well as to the communicative energies active upon the text.

2.15 From theory of binding to theory of bonding without Contradictions

Sentence parsing at a purely syntactic level is bound to very precise rules. If we move to discourse processing involving sequences of sentences, we still see collision phenomena occurring and dynamics between and among sentences. All syntactic phenomena are to be considered discourse phenomena as in Chomsky (1976, 1977, 1978). In the combination of sentences that form discourse compounds, we can observe information forces arising from changes in the arrangements of some parts of sentences. Among such phenomena are elliptical sentences, gaps, clitics and other discourse phenomena.

Bohr (1913) proposed the existence in atoms of shells occupied by a particular number of electrons; since then considerable advances have been made in understanding how energy states of atoms are related to the distribution of electrons in these shells. In the same way, syntactic patterns predictions based on the syntactic structure of a language may significantly help analyze the relationship between sentences and the information forces required to break the bonds that are present in order to create new compounds such as discourse structures are. Syntactic distribution in "sentence shells" of a given language may be strictly related to the corresponding communicative patterns expressed by a sort of "periodic table" for communicative energy detection and communicative patterns recognition in each given language.

The stability of new sentence or discourse structures is best expressed in terms of information forces and communicative energies required to break the existing bonds, forming new sentence or discourse configurations. At that point newly formed sentence structures may be analyzed as being more or less equivalent to the original structure from which they derive in terms of both pressure and temperature, that is, both in terms of communicative energy and informative force charge.

According to a theory of bonding, the study of syntax should also extend to matters of stability and polarization in sentence and discourse dynamics. This does not undermine but rather reinforces Chomsky's assessment of a unique formal abstract structure underlying linguistic expression in any given tongue.

Within Chomsky's framework the conditions for a sentence to be either well-formed or ill-formed are based on a model that provides evidence that the system of abstract rules and principles must somehow be represented in the mind of the language users.

Therefore, what we human beings "cognize" is a system of mentally represented rules and principles of a type, which a formal generative grammar intends to characterize. What each of us cognizes is a particular grammar of a specific language and this is explained by its relation to a more abstract object, a universal grammar, which is "a unified system of principles with a fairly rich deductive structure and some open parameters to be set by experience".

A universal grammar provides the basis from which knowledge of a specific grammar develops and may be thought of as a characterization not of human language but of the nature of the language faculty. A universal grammar as a biological endowment is an element of the human genotype and exposure is linked to innate properties.

What is underlined in Chomsky's framework is a biologically oriented perspective on language as a "living" system. Language is thought of as an organic system undergoing a set of transformations of different kinds. As we all know both physics and chemistry are completely compatible with biology; they are merely acting at another level of analysis.

Similarly, it is legitimate to think of the chemical composition of biological structures as merely using a different framework. This is how the CPP-TRS model differs from and complements the Chomskian model: it merely acts at a different level of analysis.

One more aspect needs to be underlined: within the newly established physical framework of language, textual phenomena may be analyzed to have their occurrence in natural language, just like organic structures. Further, textual phenomena can be created and produced by a "synthesis process" which is named "text-compression" as a kind of "inorganic compound structure". The theory of text comprehension deals with natural language organic phenomena, whereas the theory of text compression deals with textual inorganic phenomena, which are derived from organic ones.

This is a complete revolution in thinking about natural language compounds and about artificially recreated, naturally based textual phenomena.

2.16 A descriptive, explanatory and prescriptive theory of language

The nature of the theory of text comprehension is that it describes and categorizes a whole set of textual phenomena, as well as explaining it in the sense that it can justify the occurrence of new phenomena by referring to the same well-defined system of representation. The theory of text comprehension is descriptive in the sense that it embraces an infinite set of phenomena that can be subsumed under the CPP-TRS framework, but it is essentially explanatory. The theory of text comprehension is abstract enough to achieve a deep level of explanation and is able to relate phenomena that may appear unrelated.

Let us now see if in some sense it may also be considered a generative mechanism based on a commonly shared interpretative system. If there is a basic need to explain and justify the variety of textual phenomena that occur in different languages and in different communicative contexts, then the CPP-TRS component will work with a UG component and a KR component. Given the complexity of the task, it is only through a synchronous and harmonious interaction among the three that a generative model could be considered.

A whole area of research is developing to determine a predictability spectrum that explores the relationship between communicative energy and information forces in each language. It is very important to create a system of expectations about communicative patterns and textual forms in the different languages. In discussing the concept of prescriptive theory, it is wise to point out that there is a fundamental and substantial difference between the theory of text comprehension and the theory of text compression.

This new way for consistently integrating natural language as to higher communicative value of text, based on explicitating communicative function, communicative intention and communicative turn-taking visually, is actually a high transparency support based on a simple, consistent and consensual system, which directly refers to cognitive processes by reflecting them into their corresponding communicative patterns.

As in vision, texts may be recognized using communicative patterns recognition, and as in music text may be played by identifying the signs and symbols in the textual pentagram. To "comprehend" means to "understand thoroughly" and, in literary sense, to "include". Both definitions may be used to suggest that a theory of text understanding must be both accurate and precise, based as it is on very specific criteria that include various possibilities of actualization.

This is why the CPP-TRS Theory of Text Comprehension (Tonfoni 1989-94) has been conceived both as a new way to describe and visualize invisible aspects of natural language that are relevant to text processing and as a set of very precise techniques to be used in a way that is consistent with what they represent.

The distinction which was made earlier between globally and locally controlled sequences within a text has led to the definition of textual segments that result from a very specific interpretation of a naturally occurring, organic text. The study of textual segments is, of course, made possible through the introduction of the system of signs and symbols.

Such a system works not only as an annotation device but essentially constitutes a system for detection, based on analytical techniques that can be thought of as a sort of textual spectroscopy and textual chromatography.

Identification and understanding of textual segments is based on knowledge of the structure of sentences, discourse and text according to some predictable communicative patterns. All organic text segments show basic structural features that can be perceived and recognized.

Such observations about the nature of the theory of text comprehension will change dramatically when we consider and analyze inorganic text structures that are made possible within the theory of text compression.

2.17 Amplifying perception, and amplifying technology

Creating a theory intended to produce a consistent set of practical models and evolving it to a point where completely new scenarios are made visible, entails not merely the possibility of using new and powerful interpretive tools, but also the configuration of new devices and machines to perform known operations in a new and much more effective way.

An artificial system may be based on analysis and reproduction of an observed natural system, which retains its own limitations, or it may be derived from a model evolved from a natural system. A system of artificial vision may be based on the analysis of the natural human visual system with all its limitations or be based on an optimized model of vision that overcomes these limitations. Human vision depends on many structural facts. There are inhibitory processes, accommodation processes, visual acuity and different levels of adaptation.

To give an example of an active inhibitory process, we can think of the phenomenon of simultaneous contrast, where a patch of light appears to be much darker when it is surrounded by a bright background than when set against a black one. This is due to the inhibitory effect of the surrounding retina on the central region, induced by the bright surround. Many colour contrast phenomena are similarly caused.

For example, if a blue light is projected onto a large white screen, the white screen appears yellow. The blue stimulus falling on the central retina causes inhibition of blue sensibility in the periphery, hence the white background appears to be lacking its blue light. The same thing happens in communication and information processing. The distinction was made between two qualitatively different kinds of inhibitory processes: script-driven and cognitive-driven inhibitory processes.

An individual's script based reasoning may inhibit new processing and undermine knowledge. As Schank (1988) has pointed out, people tend to keep their scripts and it takes some effort and energy on their part to revise and update them. They are not always willing to do this and either consciously or unconsciously resist. Because of cognitive capability driven inhibitory processes, individuals may find difficulties in processing texts that have either too high or too low an information density. A discovery was made of phenomena of spacial summation and temporal summation occurring in text understanding as well. In other words, in spacial summation two qualitatively different kinds of textual information add their perceptive effects so that while either one alone may be inadequate to foster understanding, when presented simultaneously they may do so.

In temporal summation, two textual stimuli, which may just be words, even if each is weak, may cause understanding if presented in rapid succession within the same text. Temporal summation is consistent with a quantum theory of language; it has been shown that fluctuations in the number of quanta in a text are responsible for the varied levels of understanding; increasing the duration of a textual stimulus increases the probability of text perception.

Only a few phenomena that occur in so-called natural text perception have been mentioned. There are many more to be considered and to be taken into account when planning new communication technology that should aim toward amplifying human capabilities.

By assuming such a perspective, which is based on the earlier assumption that any new technology should amplify and expand the potential of naturally working systems, let us introduce three applications based on this as a practical example of a design of a CPP-TRS technology, starting from an optimized natural model, in which the bugs and limitation that were part of the original model have been identified. A technology that has a bug is dangerous, because this bug will be magnified and may easily proliferate. Let us start by considering the commercial application of heat or infra-red copying methods, which is thermography. "What happens in thermography is that a special kind of copy paper and the original copy are placed in contact with each other and run through a machine where they are exposed to infra-red or heat rays.

"The original copy absorbs the rays in areas darkened by print, line drawings or other illustrations, making the same impressions on the heat sensitive surface of the copy paper. The original with the new copy paper attached may be put through the machine repeatedly to produce the number of copies desired" (literal definition of "Thermography" in the Encyclopaedia Britannica).

This is the existing machine: what has been designed is an amplifier copying machine, where the original copy both undergoes a process of compression and possibly allows radiation in order to produce multiple audience and multiple target texts. This will only be possible through an accurate and consistent application of CPP-TRS scanning, CPTRU information territories definition and α, β, γ radiation. Such a tool will allow control of each set of operations performed on the text, resulting in a powerful transparency promoting technology.

Let us now consider another popular machine, the fax. "In facsimile reproducing processes, a transmitter scans the original material, converting the image into electrical signals that are sent over telephone lines or by other means of communication, such as radio or microwave. A receiver reconverts the signals into the original image on paper" (literal definition of "Facsimile" in the Encyclopedia Britannica).

This is the existing machine: what has been designed is an amplified facsimile, where the text with its annotation is sent out to facilitate translation processes based on the fact that the content is organized in such way as to make visible the qualitative interpretation of each outgoing text.

Other kinds of operations may be explained as a set of instructions for specific reformulation of precisely defined text segments.

These are just some examples introduced to provide evidence about the concept of an amplifier technology based on raising individuals' control and their monitoring of communication and information processes.

Chapter 3

The communicative positioning program and text representation systems (CPP-TRS)

3.1 A general grammar of linguistic performance

3.1.1 Introduction

In this chapter a new framework is presented and words and expressions used mostly refer to contemporary linguistics. We may therefore add that the illustration explanation and terminology certainly are chosen accordingly. CPP-TRS (Tonfoni 1989-94) is both a new model for reframing a theory of language performance as communicative action and a methodology based on a set of clearly defined conceptual tools. It also constitutes a visual language by itself, based on a set of highly articulated and consistently combinable visual elements, which are effective and practical devices for representing communicative actions.

It is based on a very clear cut assumption which is that a set of generally valid categories, playing a crucial role in linguistic performance may in fact be found and named. Generally valid categories in language are not just to be found at the syntactic level as within Chomsky's model, they may also be found at the pragmatic level. Pragmatics must be in fact be considered scientifically grounded and may be defined as the realm of "competence of performance". According to this view a text can be defined as the basic communicative linguistic entity and a new terminology is carefully introduced or redefined to ensure that new definitions will not carry old meanings and that redefined terminology will not cause any ambiguity. New concepts such as communicative positioning program have been carefully explained, redefined and reframed. It is equally necessary that existing concepts such as linguistic performance are analyzed so they can be agreed and mutually shared concepts and practical interpretative tools.

3.1.2 CPP-TRS: a new framework

In order to explain why a paradigm that is explicitly intended to address old questions and solve old problems in new ways is both consistent and really new. It is very important to make sure that definitions and terms are mutually understood. First, let us go back to the difference between "competence" and "performance" (Chomsky 1957, 1965). Let us add that a new perspective on language as communicative action attempts both to describe and explain what a competent performance is. In Chomsky (1965) competence is "the speaker's-hearer's knowledge of the language" contrasted with performance, which is "the actual use of language in concrete situations". What needs to be added, however, is that any new approach to performance needs to be scientific and rigorous; not merely observation driven, but explanatory and predictive as well.

The current approach differs from others primarily because of the basic problems identified as to be the relevant ones. These may be synthesized in the following way:

There is something basic in language, which may be found not solely at the syntactic level but also at the pragmatic level. There is a generally valid set of rules but these are not merely syntactic rules. They are communicative positioning attitudes that we can find in any language. They work on top of the different languages and all languages reflect them in a way that is so consistent that they may be identified, in spite of the differences among languages and the various culture-specific components, by language-dependent patterns that allow native speakers to recognize them.

There are generally valid categories, like syntactic categories, which are defined as communicative positioning attitudes, which are consistently reflected in texts and text segments in the form of communicative patterns, so that the native speakers may recognize them as active while they are themselves carrying on a conversation. The basic entity for communicative positioning detection and identification is not just the single sentence but the entire text, as sentences cannot be properly handled if extracted out of their own originating flow.

Sentences which we may also refer to as text segments may show an alternation of different communicative positioning attitudes, as while expressing ourselves we do in fact vary our style continuously and very naturally. Native speakers have an immediate capacity to discriminate between what may be considered to be a well formed sentence such as "I like apples" from an ill formed sentence like "Like I apples" at a syntactical level, which is true not just for sentences, but also for more extended texts.

Native speakers have a particular competence for deciding whether a sentence is acceptable, correct or incorrect; but they equally have a communicative competence as to discriminate between pragmatically correct sequences and pragmatically incorrect ones. So if a child is to ask his mother "mom would you give me some apples now?" this is found to be consistent whereas the same request would not be consider pragmatically correct if the child would address one of his mother friends using the same sentence.

"Rule governed creativity" not only justifies an infinite number of sentences as being the output of a definite set of rules, but also justifies an unlimited set of texts, which consistently represent a limited set of communicative positioning attitudes.

Native speakers have an innate competence and capacity to produce and interpret an infinite number of sentences, to evaluate their syntactic correctness and to notice special relationships between them, such as an active versus a passive sentence, but native speakers and native writers are also competent enough as to perceive an infinite number of texts and text segments and evaluate them against their respective communicative positioning and pragmatic correctness.

Special relationships between informative versus explanatory texts, and analytic vs synthetic texts are particularly evident and very much perceived.

Both introspection and observation of syntactic phenomena are needed, so is self-reflection on communicative positioning and cognitive awareness of other individuals' communicative positioning, in order to ensure that communication may really work effectively and no misunderstanding may occur. If we consider the notion of 'correctness', we can see it as dependent on syntactic criteria. Syntactic criteria are being part of a biologically determined and innate background of knowledge as for Chomsky.

According to the communicative positioning perspective the notion of 'correctness' is redefined in terms of consistency and is related to cognitive criteria, given a certain set of communicative attitudes that are part of a cognitive program. We can thus say that communicative positioning attitudes are consistent with the cognitive structure we individual human being have and activate in a certain culture and language. We can also add that the result of a certain communicative positioning attitude or attitudes needs to be consistent, in the sense of reflecting such an attitude or attitudes as it will then show in text which is in fact the end result. If there is a discrepancy between the communicative positioning attitude intended and its reflection on text, then problems in interpretation and actual understanding may occur.

A real "competence of performance" shown by native speakers exclusively as for their own language and culture should prevent such discrepancy from occurring, unless it is intentionally kept of course. Some discrepancy may occur in active speakers if cheating or just uncooperative. As communication is based on cooperation and mutual sharing of an agreed and accepted system, a discrepancy between a communicative positioning attitude and its reflection on a text or part of a text, may be viewed like a syntactic rule and its violation in a certain sentence.

If native speakers are able, based on their competence in their native language, to recognize, name and categorize different elements of a sentence as subject or article or verb, noun phrase and verb phrase, the same individuals should be able to recognize, name and categorize differently positioned texts and text segments as representing differing positioning attitudes.

Just as transformations occur when applying the system of syntactic rules (Chomsky 1957), parts of the same text may undergo a set of transformations even unintentionally as for a joke, which may easily be turned and taken as a real assessment. Consistency between the communicative positioning level and the respective text representation level, both as for local paragraphs or global documents, is what matters to ensure that the document may in fact be processed as it was intended, consistency being the result of real accurate matching.

A text may then be also considered to be consistent within itself, holding together and retaining form, as a result of this. The problem of inconsistency between language and intention has been widely discussed in linguistics (Austin 1962; Searle 1968). The speech act theory (Searle 1968) identifies a set of verbal categories representing intentions. Along those same lines some specific terms and verbs have been technically redefined within the new framework here introduced as to represent specific communicative intentions, which then determine a certain stylistic choice, which then reflects upon sentences and paragraphs.

These terms have been turned into interpretative cues and categories, which apply to text and texts segments.

Defining precisely what a certain communicative positioning attitude may entail or imply is in fact of extreme relevance not to expand the number of categories to the extent that the system becomes unmanageable and ineffective. On the other hand, it is important to make sure that each of the categories introduced may in fact cover all varieties of text occurrences, which are indeed unlimited. This is precisely what happens in syntax, as selected categories need to be adequate as to cover all possible sentence outputs. The number of categories need to be restricted, otherwise may the system become unmanageable.

What we mean to say by specifying that the system is in fact cognitively bound is that there is indeed a common cognitive structure and program, that allows any individual speaker of a certain language in a certain cultural context to recognize a set of communicative positioning attitudes, which definitely may generate a quite unlimited set of text occurrences. What we mean to say by culturally bound, on the other hand, is a common set of assumptions and beliefs from a certain culture which reflect upon language and communicative performance as for a set of agreed upon and context-sensitive rules.

In other words there is a special sensitivity, as for native speakers, which allows any individual out of a certain culture and context, to trace back a certain communicative behaviour and communicative pattern to its originating cause and therefore consistent meaning attribution. These two aspects, cognition and culture, are intrinsically linked, but also must be analysed as being entirely different components working at different levels.

Research has been pursued and carried on in knowledge representation especially as for how some kind of knowledge is organized around certain facts and events at various levels of complexity and as part of some cultures and cultural assumptions.

A whole set of models in knowledge organization have been designed to be used for knowledge filing, retrieving and indexing, such as the frame (Minsky 1975), the script (Schank 1977), the plan (Wilensky 1983), and the plot unit (Lehnert 1981). All of these models address the problem of knowledge organization in certain domains and for specific purposes and as part of a certain culture. Analyzing culturally bound communicative endeavours poses the need for accurate observation about very specific facts and events.

In order to be able to correctly identify and consistently recognize and label cognitively bound communicative attitudes, self-reflection and cognitive awareness is strictly needed especially on those who know a language and a culture some but not as if they were in fact native speakers. Self-reflection is an in depth discovery process intended to enlighten by making them first surface a set of cognitive processes, reflecting upon communicative positioning attitudes, which then reflect upon stylistic choices, which we may then see occur in a final text both as for sentences or entire and more extended sequences of sentences.

Cognitive awareness, being the result of self-reflection, generates the capacity to recognize other individuals' communicative positioning attitudes from the way they are reflected on texts or text segments. Cognitive awareness is to be considered as a purely human capacity which no automatic system may incorporate the added value, represented by active and protagonistic role by human interpreters as for documentation processing. In order to proceed scientifically in a realm where many components are invisible as in the realm of language and communication, an important step, in addition to the accurate definition of terms, is visualization.

Visualizing is somewhat explaining a complex issue or phenomenon and making it transportable throughout a set of languages and cultures. Visualizing invisible processes or elements has always been a very important and common practice in the evolution of science and scientific models. It is therefore a tremendous support for enhancing self-observation capacities and cognitive awareness.

If a certain explanatory model is found to be cognitively consistent, and also easy to see as visualized consistently, it is very likely to become accepted and will also be used in different contexts and languages.

Along the same lines a cognitively consistent methodology, which means also a way to operate as for solving a certain set of problems, may be used in many and diverse languages. If it is in fact cognitively bound it means that those very same invisible aspects and processes identified are active in all languages though in a diverse set of combinations. By having them somewhat surface may these invisible processes all be defined accurately and they may also be complemented with a consistent visual representation. Once each single process has been explained extensively and in depth too, may this explanation become captured and miniaturized in a visual representation which is meant to be highly synthetical by nature.

Communicative positioning attitudes have been therefore defined very carefully by providing and attaching a specific meaning to a set of ordinary terms, which have therefore been turned into highly technical terms.

Each of these terms may in fact be supported or substituted by an equivalent visual representation, which can be used as to represent visually what the term means. This is why the whole framework here illustrated is based on a precise terminology, which carries its own visual representation along, articulated into a whole and highly diverse set of visual dynamic schemes, such as textual machines and textual canvases, meant to represent the deep structure of text in the process of its creation, and a set of icons, articulated into global signs, local symbols and turn-taking symbols meant to annotate text in its final stage which means when finally fully consolidated.

As a cognitively consistent visual system to be used on top of language, it may be used and adapted to a whole variety of languages. There are many situations where invisible aspects and processes which constitute the communicative positioning of a document may most productively be visually integrated within the document itself as to preserve its intentional meaning and originating context. As to see some continuity with an already well established research tradition, we may certainly add that visualizing, defining and creating articulated metaphors to explain complex phenomena or processes, which cannot be seen, is often practiced in a very recent scientific framework, as the one of artificial intelligence.

In "Society of the Mind" (Minsky 1986) the metaphor of a society is used to represent the entire functioning of the mind in its various components. Some new terminology is introduced complementing a really large set of redefined terms, which represent the explanatory core of the theory.

The positioning framework here introduced is intended to first define and then provide a generally valid grammar of communicative positioning actions, which may in fact constitute the core grammar of that competence of performance human beings do have. In order to be considered as scientifically grounded, any theory of language should entail observation, explanation, description and definition in order to make predictions and evaluations based on a set of clearly identified criteria possible in the first place. A first set of visual and dynamic schemes meant to represent the process of creation in text is now introduced. They are specifically textual canvases and textual machines.

3.1.3 Deep structure of texture

A “canvas”, represent a specific space for painting to be progressively filled in with drawings and then colors. A textual canvas may metaphorically represent an overall “deep structure” or defined and opened space for an a pre-identified communicative plan to be then implemented in full details.

A “machine” is a device for energy transformation and transmission. A textual machine is meant to physically model a whole set of processes which may not be easily explained just in words as they are quite complex and abstract. A set of dynamic representation systems as textual machines are is meant to promote a learning by examples and by actually seeing. Textual machines are devices for observing and defining what has not up to now been observed or defined.

Textual machines are intended to represent visually, physically and dynamically different kinds of transformations occurring in text at the deep structure level, which means where and when major choices as for the overall text project and design are made. Textual canvases and textual machines are intimately connected in that they both relate to the “deep structure”. As such, they are extremely useful tools for self-reflection and self-observation.

3.1.4 Textual canvases

a. Density in texture

This canvas, illustrated in Fig. 3.1, shows the different possible densities of information, which may be chosen as to design and plan a text accordingly.

Figure 3.1. Example of density in textures

b. Word explosion and word chaining

This canvas, illustrated in Fig. 3.2, represents the cognitive process of free association triggered by a word. It also depicts how a group of words can be linked together after some brainstorming has occurred, in the form of an explosion, so that they form a logical, linear chain. The word chain that results from this process will first be a sentence but can also be extended to a set of sentences and paragraphs.

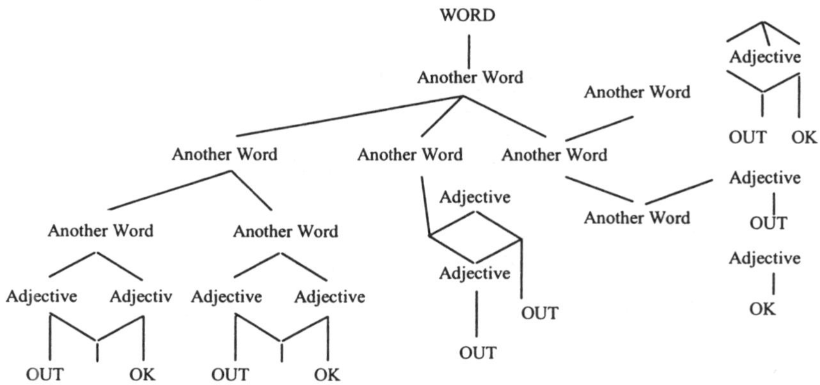

Figure 3.2. Example of word explosion and word chaining

c. Metaphor creation

This canvas, illustrated in Fig. 3.3, shows how a metaphor may be generated out of a two words explosion: one for the source word and one for the target word. The source word is the word we start with, and the target word is one that has been chosen for transferring a new meaning to the source word.

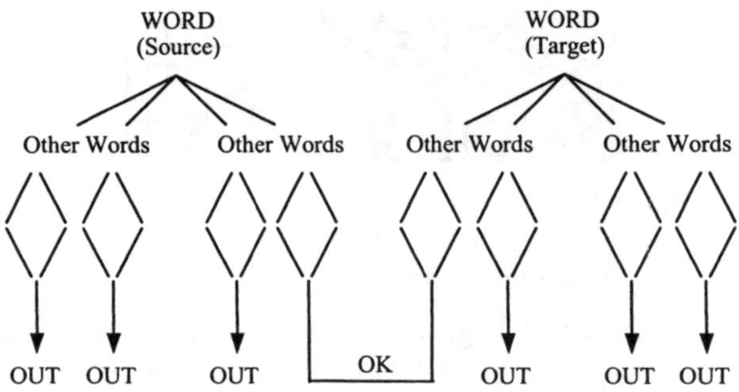

Figure 3.3. Example of metaphor creation

d. Pattern recognition and analogies in stories

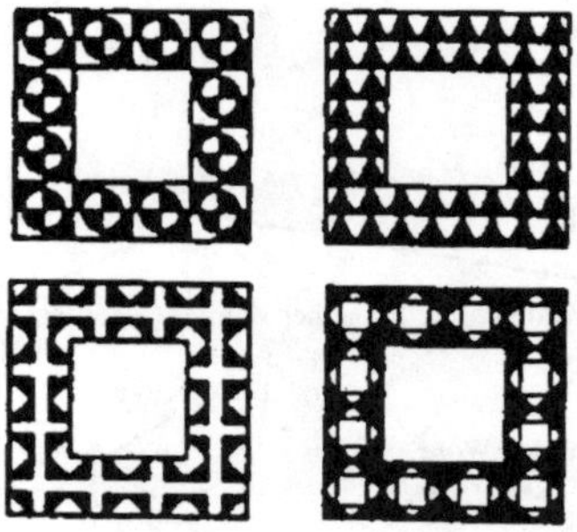

Figure 3.4. Example of pattern recognition and analogies in stories

This canvas, illustrated in Fig. 3.4, shows how a visual frame may represent the logical structure of a story at a conceptual level. Once a story is visually represented, it becomes easy to replicate its structural patterns as to create other conceptually analogous stories.

e. Global narrative text planning

This canvas, illustrated in Fig. 3.5, represents a visual scheme for globally planning of a narrative text. This canvas enables us to visually represent a long story that may encompass a number of events, with different protagonists and many possible outcomes. It opens up the possibility of conceiving many different ways to develop the story while retaining and checking its coherence.

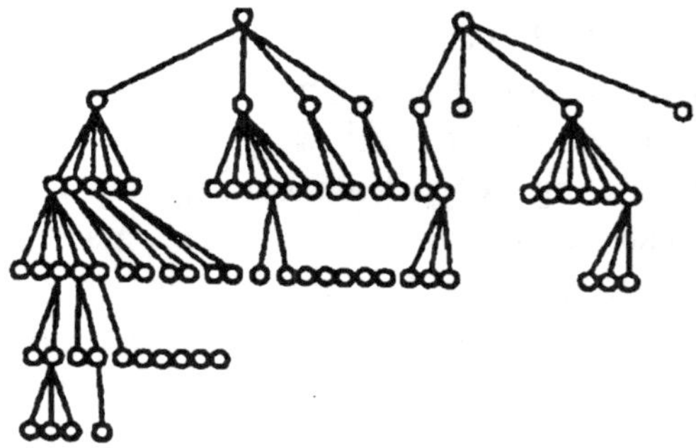

Figure 3.5. Example of global narrative text planning

f. Local narrative text planning

This five-part canvas, illustrated in Fig. 3.6, helps us visualize the details of a narrative text. It visually indicates ways to represent backward and forward links in a narration, to visualize a story in different ways from a particular point, to use temporal and causal connectors. It is meant to facilitate establishing links as for narrative concepts, and to reconstruct the story differently by changing its sequence.

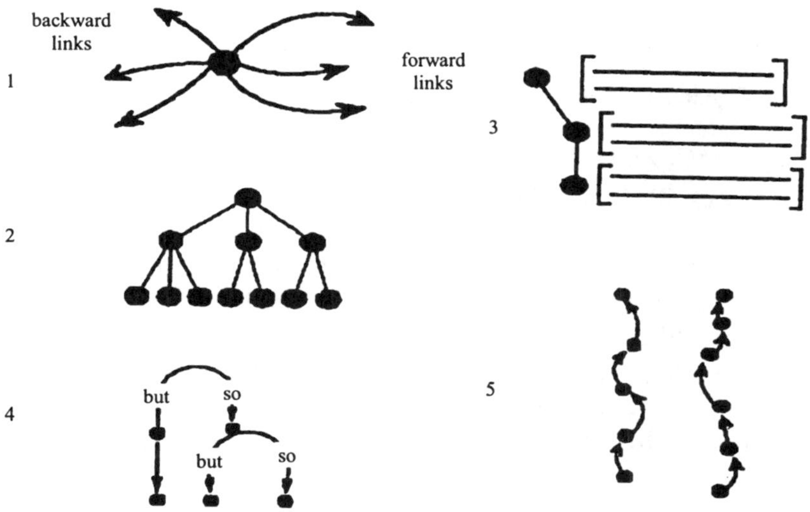

Figure 3.6. Example of local narrative text planning

g. *Analogous story planning*

This canvas, illustrated in Fig. 3.7, is a visual aid for planning and then generating a whole set of different stories, which are analogous to a particular and originating story. It is meant to help visualize the different ways that two stories are analogous. Stories may in fact be partially analogous- which means they are initially, intermediately, or finally analogous- or they can be totally or globally analogous.

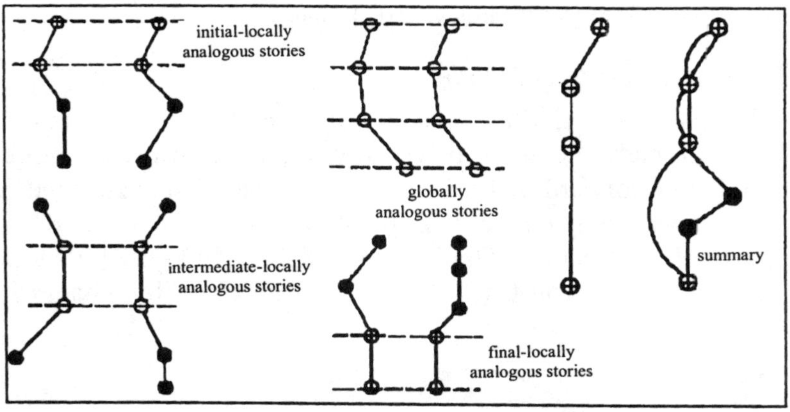

Figure 3.7. Example of analogous story planning

h. *Explanatory text drawing*

This canvas, illustrated in Fig. 3.8, provides us with a set of visual schemes for planning and organizing an explanatory text by exhibiting a logical progression or events. The first visual scheme resembles a traditional outline and shows the logical unfolding of the text, starting with a title and flowing down through the division and development of a set of paragraphs. The second scheme provides an alternative to the strictly logical progression of paragraphs, and rather focuses on identifying and maintaining a topic all the way throughout the text.

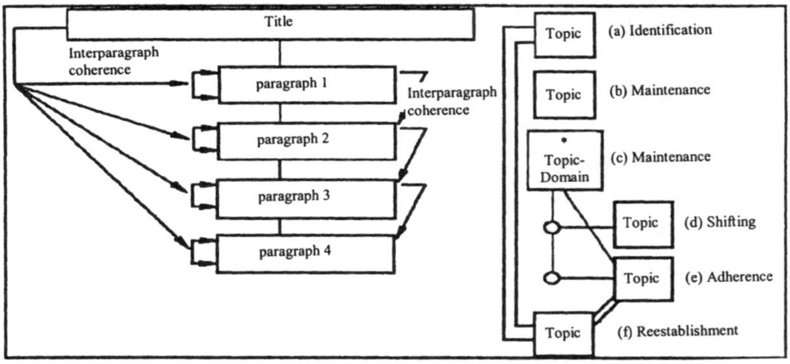

Figure 3.8. Example of explanatory text drawing

i. Dialogue-conversation

This canvas, illustrated in Fig. 3.9, is for visualizing a conversation, recording its rhythm, and tracking its evolution. It is divided by a vertical line into two parts: the left one represents and retains the contributions of the initiator of the conversation, while the right one represent and retains those of the addressee. The lines flowing down through both parties involved are meant to trace the dynamics of the dialogue as it progresses. The process of planning and drawing a conversation on a canvas has the effect of slowing it down so that conversation partners may stay focused and head towards the same goal. This canvas also enables a detailed, written reconstruction of the conversation.

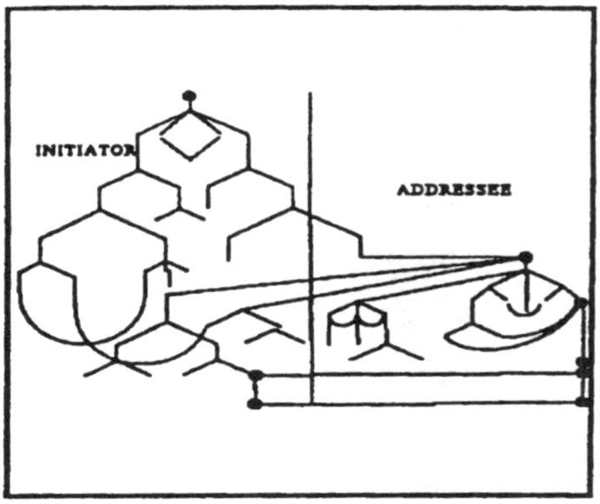

Figure 3.9. Example of dialogue-conversation

j. Collective text perception

This canvas, illustrated in Fig. 3.10, is a two-part canvas. The first part helps us visualize a particular type of collective text planning called "collage writing". Collage writing may be performed by a group of individuals who collectively endeavour to create a text by piecing together their individual written contributions. The second part of the canvas represents "mosaic reading". Mosaic reading shows how different readers may read the same text in different ways. In mosaic reading, a small group of people may in fact summarize the same text and then compare them to see how their interpretation and selection processes differ.

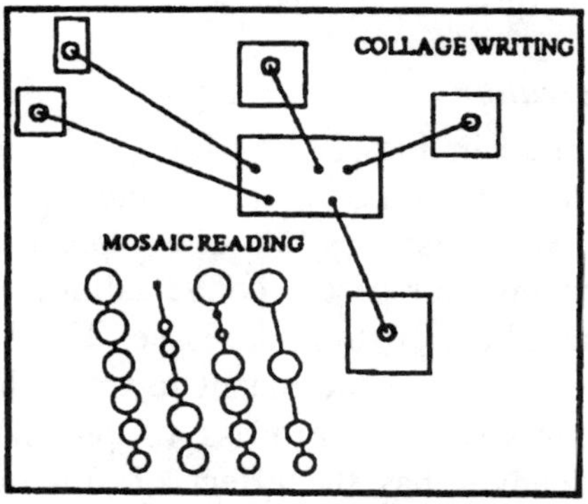

Figure 3.10. Example of collective text perception

k. Text rotation

This canvas, illustrated in Fig. 3.11, visualizes the reformulation and content reorganization process as for different texts which may carry the "same" text to a diverse readership. Different readers who have differing information needs will be provided with a different kind of text in order to accommodate their particular perspectives and expectations still keeping pretty much the same content structure.

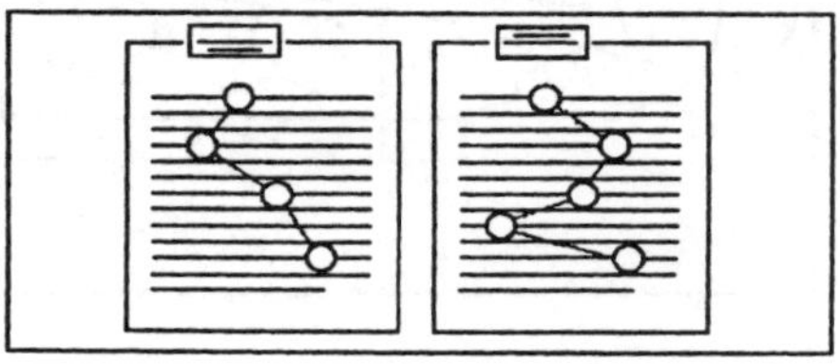

Figure 3.11. Example of text rotation

l. Hypertext

This canvas, illustrated in Fig. 3.12, is particularly suited for designing "hands-on" explanatory texts such as instruction manuals or procedural guides. This is non-linear and intended to be used to fulfill differing information needs.

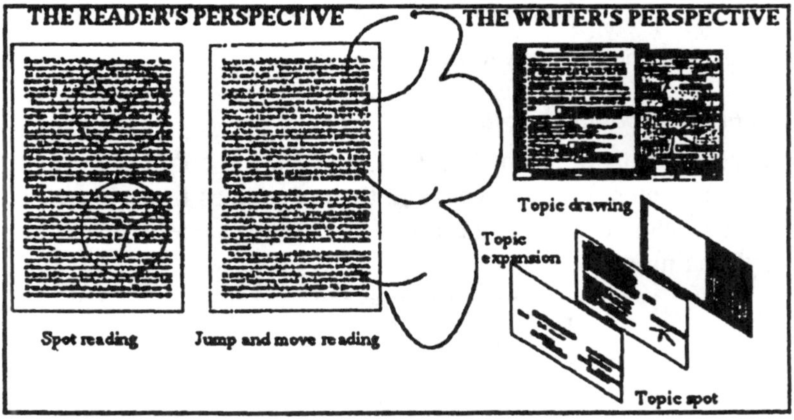

Figure 3.12. Example of hypertext

To conclude, designing a text needs to be subject to a previous analysis as for readership and scope of course and textual canvases here illustrated may be considered both as practical tools for writing and also as interpretation devices for communicative positioning detection out of reading other texts as to retain information.

3.1.5 Textual machines

Textual machines are dynamic visualization tools for representing different cognitive processes active both separately and together. There are ways to think metaphorically and represent analogically what is in fact happening while we are writing and reading. Let us explore their meaning.

a. Machine for simple rotation of a word

This machine, illustrated in Fig. 3.13, models the cognitive process of freely associating words with one particular word. It is a three-dimensional representation of the same concept as the canvas for word explosion.

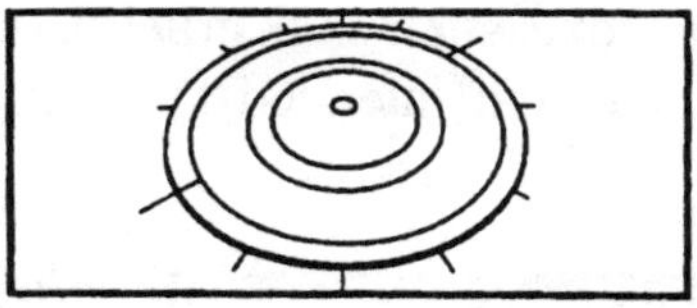

Figure 3.13. Machine for simple rotation of a word

b. Machine for complex rotation of a word

This machine, illustrated in Fig. 3.14, models the cognitive process of selecting attributes or adjectives, and then organizes them linearly. It is a three-dimensional representation of the same concept as the canvas for word-chaining.

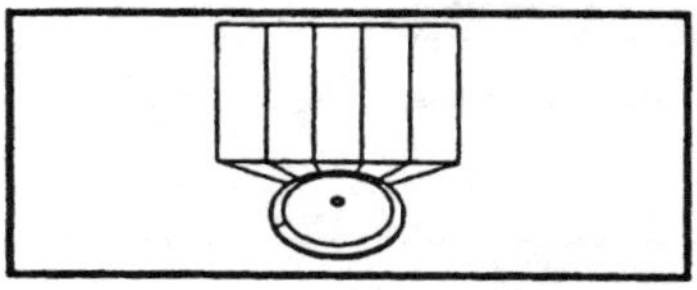

Figure 3.14. Machine for complex rotation of a word

c. Text motion machine

This machine, illustrated in Fig. 3.15, provides a dynamic, three-dimensional way to think about a narrative text in which many possibilities are kept open. It is a three-dimensional representation of the same concept as the canvas for planning a narrative text.

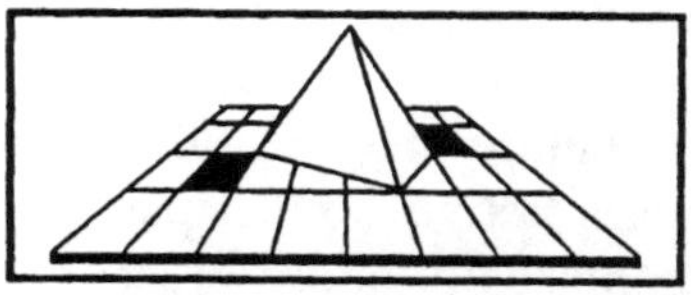

Figure 3.15. Text motion machine

d. Textual lens

This machine, illustrated in Fig. 3.16, models the cognitive processes of focusing on some part of the text and then expanding the aspect that is considered to be most relevant. It is a three-dimensional representation of the same concept as the canvas for textual rotation.

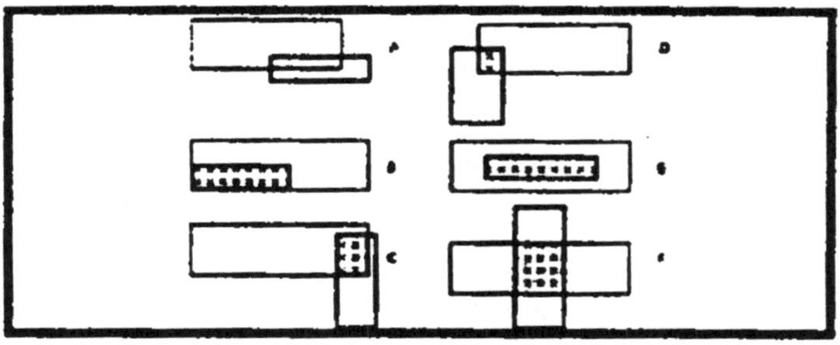

Figure 3.16. Textual lens

e. Wheel for text browsing

This machine, illustrated in Fig. 3.17, models the way information is processed and how much the text implicitly or explicitly conveys, and how narrative text has forward and backward links. It is a three-dimensional representation of the same concept as the canvas for planning narrative text in a detailed way.

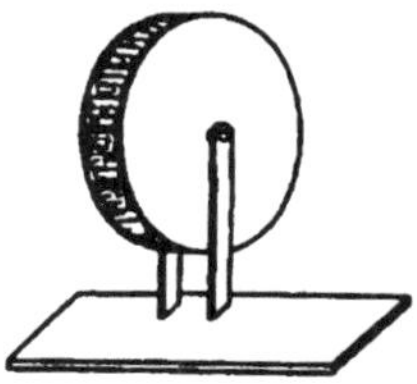

Figure 3.17. Wheel for text browsing

f. Carriage for memory text

This machine, illustrated in Fig. 3.18, enables us to see the relationship between an evocative word and a memorized story. It demonstrates how the same word may trigger many different stories. It is a three-dimensional representation of the same concept as the canvas for finding analogies in stories.

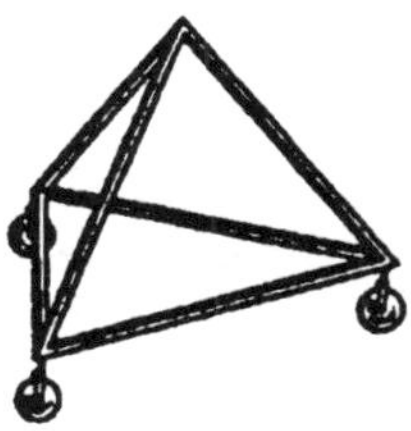

Figure 3.18. Carriage for memory text

g. Carriage for explanatory text (simple)

This machine, illustrated in Fig. 3.19, visualizes the process of knowledge acquisition and storage. It is a three-dimensional repre sentation of the same concept as the canvas for planning and drawing explanatory texts.

Figure 3.19. Carriage for explanatory text (simple)

h. Carriage for explanatory text (complex)

This machine, illustrated in Fig. 3.20, is a more complex version of the machine above. It models the process of knowledge acquisition, storage and retrieval. It is a three-dimensional representation of the same concept as the canvas for hypertext.

Figure 3.20. Carriage for explanatory text (complex)

i. Three-floor rotation machine for poetry

This machine, illustrated in Fig. 3.21, models the structuring of evocative text that triggers emotions and personal feelings. It is a three-dimensional representation of the same concept as the canvas for creating a metaphor.

j. Machine for transferring text

This machine, illustrated in Fig. 3.22, models how a text constructor can organize the same topic in two different text styles. It is a three-dimensional representation of the same concept as the canvas for organizing the density of the text.

k. Device for transforming text in phases

This device, illustrated in Fig. 3.23, models the process of transforming text paragraph by paragraph from some initial version to a final version. It is a three-dimensional representation of the same concept as the canvas for organizing the density of the text.

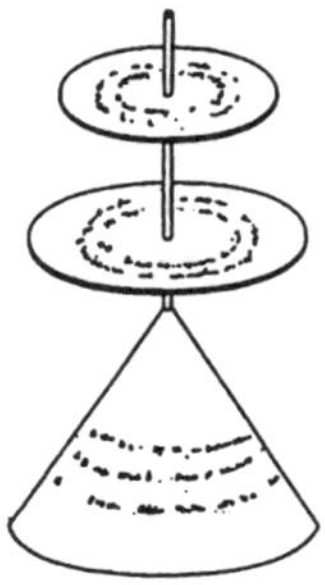

Figure 3.21. Three-floor rotation machinery for poetry

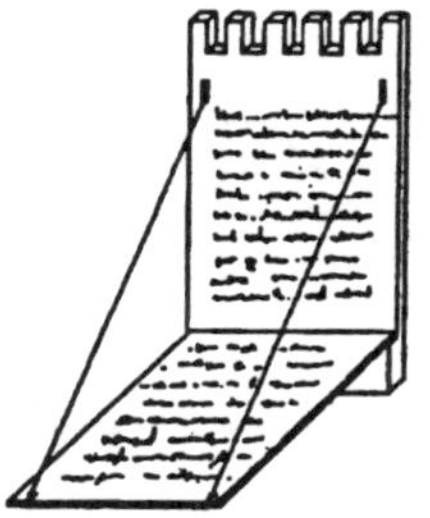

Figure 3.22. Machine for transferring text

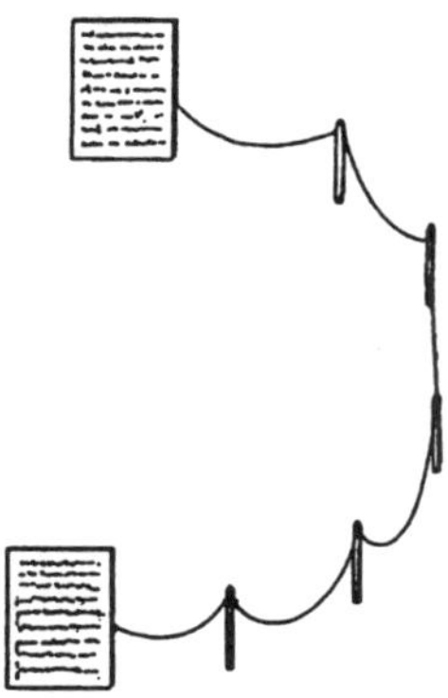

Figure 3.23. Device for transforming text in phases

l. Focusing machine

This machine, illustrated in Fig. 3.24, helps us to understand the importance of focusing as it demonstrates how different individuals focus differently on the same text. It is a three-dimensional representation of the same concept as canvas for mosaic reading.

m. Machine for producing ideas and turning them into text

This complex machine, illustrated in Fig. 3.25, models how one or more evocative words initiate the processes of planning and organizing a text. It is a three-dimensional representation of concepts that have already been explored, as in word-explosion, focusing and linearization.

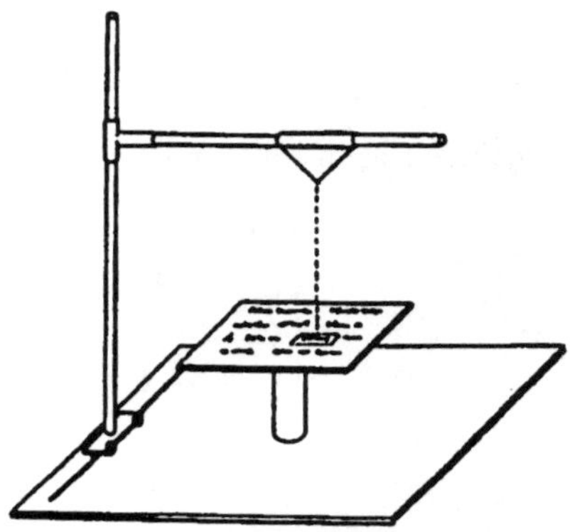

Figure 3.24. Focusing machine

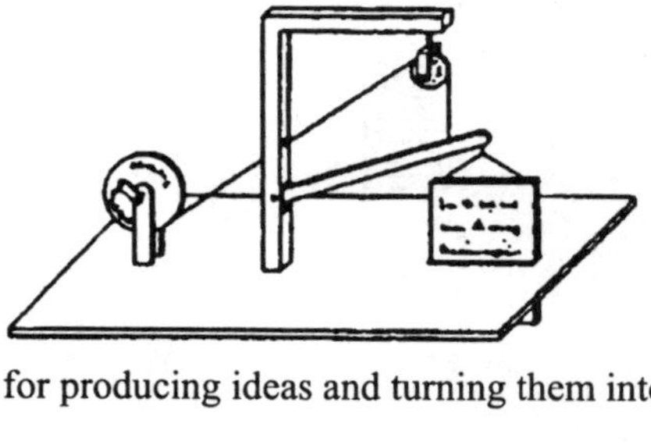

Figure 3.25. Machine for producing ideas and turning them into text

n. Textual catapult

This machine, illustrated in Fig. 3.26, enables us to identify a missing element from the text that can be added, based on the context. The context is the expectation that the text itself creates. This is a three-dimensional representation of a whole set of concepts that have already been explored as in text planning and implicit versus explicit knowledge recognition.

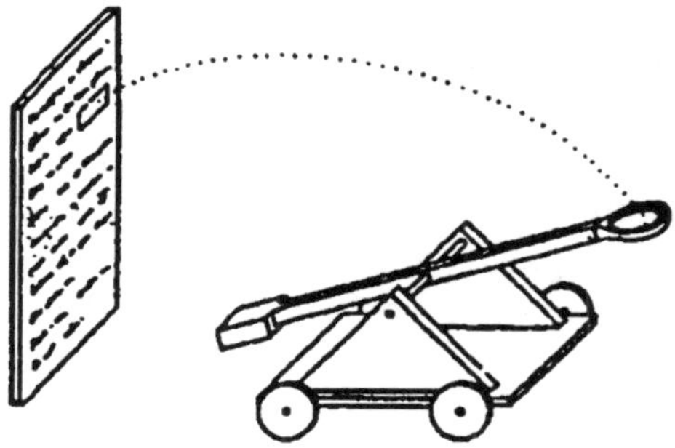

Figure 3.26. Textual catapult

As a result of many processes occurring while writing and reading, some quite stable and highly recognizable positioning patterns may easily emerge. The theoretical model presented here does in fact also provided a methodology for recognizing communicative positioning attitudes.

3.1.6 Surface structure of a text

Let us now consider what may be observed at the surface level of each text and let us more precisely focus on written texts. After having decided about a certain plan for a given text to be, here called texture, writers must choose and clearly define their final choices. This will not prevent them from changing again later by taking an already consolidated text or paragraph or a single sentence as to reformulate it either entirely or partially. Such process may also be called more generally stylistic conversion.

Let us now go back to the issue of writing meaning here actualizing text and stylistic choices, in a way which may adequately reflect communicative positioning attitudes which have been adopted and need somehow to become visible and surface in the overall surface structure of text and also as for each single paragraph.

What this new way of looking at language and of visualizing text seeks precisely to describe is the speaker and writer's as well as the hearer and reader's competence of performance, which resides in the capacity to generate and recognize communicatively positioned texts as for what they were intended to be.

Rephrasing Radford (1988), we are considering here a sort of mental property, that may have two physical manifestations, written or spoken language.

The existence of a cognitive structure, which makes it possible for us to understand each other at that level of communicative positioning behind and beyond single words and sentences is therefore the basic assumption. Both written and spoken forms of language reflect such cognitive structure based performances. Thus linguists will not merely have to confront the complexity of spoken or written language; as they will be able to reach out as for the in depth and in breadth study of language in all its complex and multiple dimensions.

Those different components which together determine the communicative positioning of a text and its parts may be represented in two comprehensive schemes. First come synthetical models, which apply to the deep structure of texture and then come icons which apply to the surface structure of a text as for its paragraphs and sentences.

Let' s therefore carefully analyse the following schemes, which are Fig. 3.27 and Fig. 3.28 respectively.

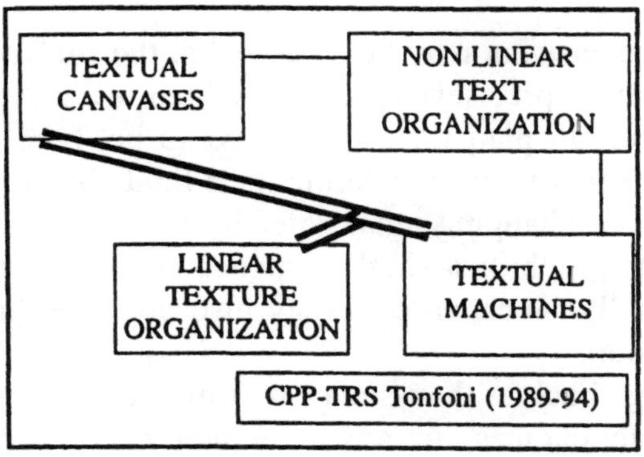

Figure 3.27. Deep structure level of texture

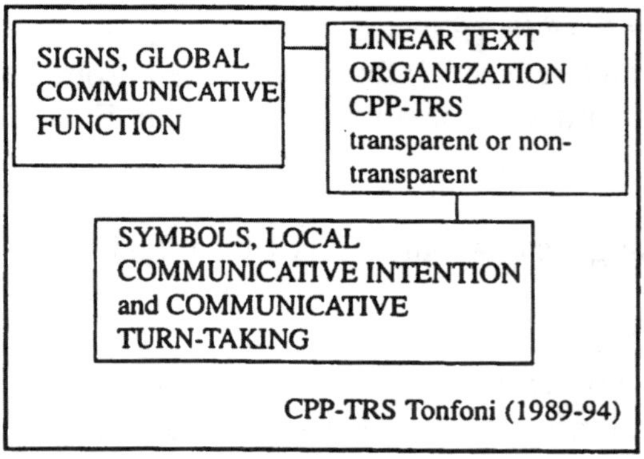

Figure 3.28. Surface structure of text

Let us proceed further as to illustrate the different components that carry specific cues for representing communicative positioning attitudes at the surface level of text, which is in the actualized text as for paragraphs and sentences. More specifically signs are more global control icons that indicate the communicative function and type of a full text. Symbols are more local control icons that indicate the communicative intention of each paragraph and sentence.

Turn-taking symbols may be applied both globally and locally and are icons that indicate the role the communicative receiver should play. Signs and symbols may all be combined to define accurately the many and complex dimensions communicative positioning entails. Signs and symbols are immediate and non-ambiguous; they may be combined and thereby constitute a visual system for communicative positioning attitudes representation.

3.1.7 Textual signs

Textual signs indicate that a paragraph or sentence must be interpreted as directly referring to a specific global communicative function controlling it.

a. Square

For general information and general story (illustrated in Fig. 3.29).

b. Square within a square

For summary and reduced information taken from a previously provided general information or story (illustrated in Fig. 3.30).

c. Frame

For a story analogous to another story (illustrated in Fig. 3.31).

d. Triangle

For a memory text and subjective experience triggered by previously provided general information or story (illustrated in Fig. 3.32).

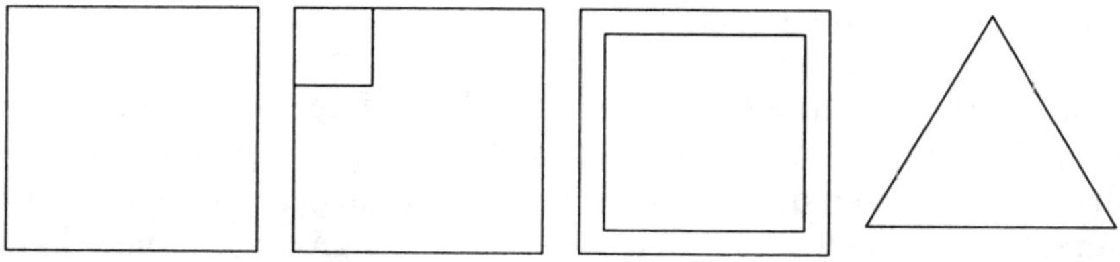

From Left to right:
Figure 3.29. Square
Figure 3.30. Square within a square
Figure 3.31. Frame
Figure 3.32. Triangle

e. Circle

For a global or general concept that emerges from a previously provided general information or story (illustrated in Fig. 3.33).

f. Grouped semicircles

For concepts that emerge from previously provided general information or story (illustrated in Fig. 3.34).

g. Semicircle

For a particular concept extracted from previously provided general information or story (illustrated in Fig. 3.35).

h. Inscribed arcs

For adding an alternative to a story or previously provided general information (illustrated in Fig. 3.36).

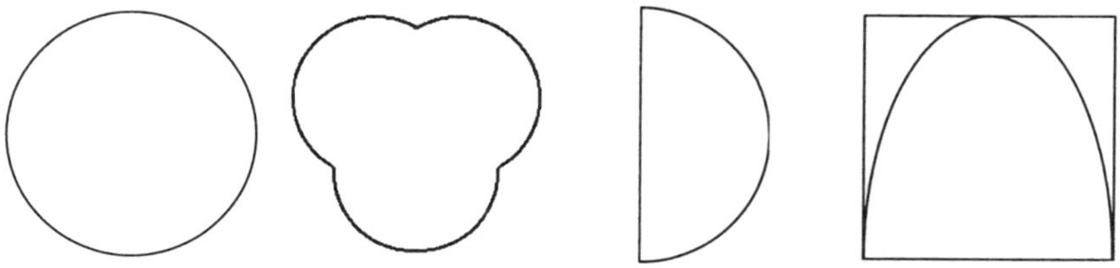

From Left to right:
Figure 3.33. Circle
Figure 3.34. Grouped semicircles
Figure 3.35. Semicircle
Figure 3.36. Inscribed arcs

i. *Opened text space*

For opening up a new text segment in a text for an alternative to a story or previously provided general information (illustrated in Fig. 3.37).

j. *Right triangle*

For comment deriving from previously provided general information or story (illustrated in Fig. 3.38).

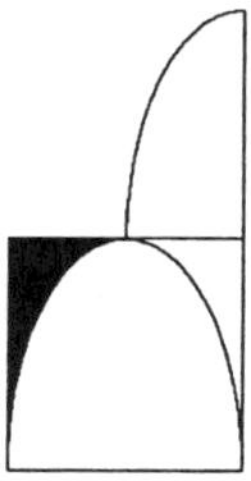

Figure 3.37. Opened text space

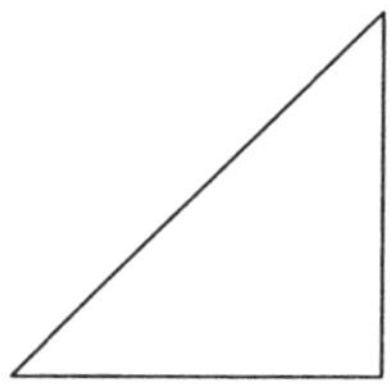

Figure 3.38. Right triangle

3.1.8 Textual symbols

Textual symbols indicate that a paragraph or sentence is controlled by a specific communicative intention, which is active at a local level.

a. *Describe*

From the Latin "describo", to write around. Meaning: the provision of as much information as the author wishes, starting from a point at the middle of the spiral and proceeding by expanding the spiral; there is no need to follow a logical-chronological order; the spiral may be small or large, depending on how much information is being presented (illustrated in Fig. 3.39).

Figure 3.39. Describe

Figure 3.40. Define

b. Define

From the Latin "definio", to put limits. Meaning: the provision of only relevant information about a certain point, indicated by the middle point of the square and the limit drawn around it; there is no need to follow any logical-chronological order but one must select and choose the relevant information to be kept within the square. "Define" is "describe" under certain constraints; in a dictionary, the definition of a word or a concept implies that one must choose what is relevant to be written in order to understand the specific meaning (illustrated in Fig. 3.40).

c. Narrate

Meaning: the presentation of facts and events by following a logical and chronological order; there are points following each other on a line. The longer the story is, the more points you may add. Arrows indicate the logical reading direction of facts and events. If the story becomes longer, more points are added on the right side and the line becomes longer (illustrated in Fig. 3.41).

Figure 3.41. Narrate

d. Point out

Meaning: the extraction of a point, specific event or fact, from a narrative (story), focusing on that, adding more information and thereby expanding it. The visual symbol represents the specific point that has been chosen and expanded within a chain of facts and events. Within a story it may represent the need for further information and details about a part of it (illustrated in Fig. 3.42).

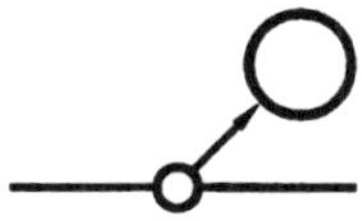

Figure 3.42. Point out

e. Explain

From the Latin "explano", to unwrap, to open up. Meaning: the representation of facts in a cause-and-effect order. It is possible to start from the original cause and descend progressively to a set of effects or move from effects to the original cause. This is why the symbol, a graphic representation of the procedure, works in both directions: from top to bottom and bottom to top (illustrated in Fig. 3.43).

f. Regress

From the Latin "regredior", to go back. Meaning: regression here means giving more information about a certain item within a chain of information. The visual symbol represents the process of focusing on an item and the process of depth information expansion, which is activated just above that point. You may have different levels of regression according to differing information needs (illustrated in Fig. 3.44).

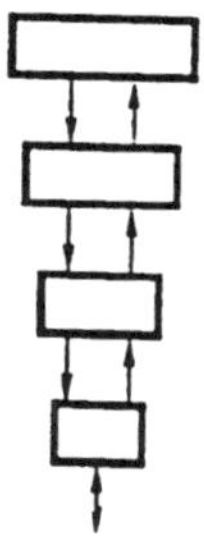

Figure 3.43. Explain

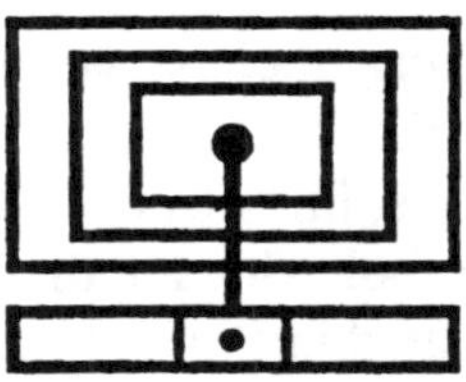

Figure 3.44. Regress

g. Inform

From the Latin "informo", to put into shape, to shape up. Meaning: packaging of information. The visual symbol is double-faced: you can read it from left to right or from right to left (illustrated in Fig. 3.45).

Because of their strong visual impact, symbols (1) and (2) are used to represent the concepts of synthesis and analysis in general and may be added to the other symbols as well (define, describe, explain, etc.).

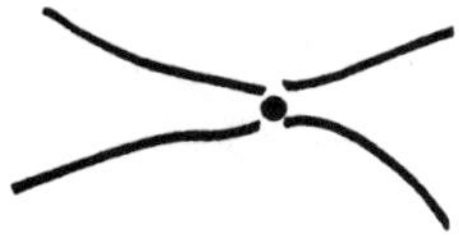

Figure 3.45. Inform

g.(1) Inform synthetically

First Phase. From large to small: synthesis: moving from a large bank of information towards a synthesis (central point) (illustrated in Fig. 3.46).

g.(2) Inform analytically

Second Phase. From large to small: analysis: moving from a limited set of data (central point) towards more analytical data (illustrated in Fig. 3.47).

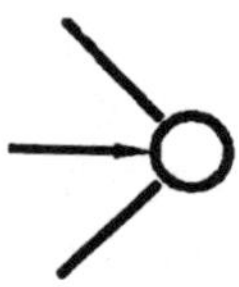

Figure 3.46. Inform Synthetically

Figure 3.47. Inform Analytically

h. Reformulate

From the Latin "reformo, reformulo", to change shape and to shape again. Meaning: the change of attitude in information packaging: moving from explanation to description or definition or information, etc. In practical terms, this means substituting a cognitive attitude with another, a visual symbol with another. This symbol means "change", "substitute" (illustrated in Fig. 3.48).

Figure 3.48. Reformulate

Figure 3.49. Express

i. Express

From the Latin "exprimo", to push out, to press out. Meaning: the provision of personal and emotional opinions about facts. This is the most subjective mode of information packaging and is bound to personal experiences such as "I like / I dislike". The visual symbol represents a visual territory which is open and subject to modification according to personal opinions: an amoeba - like shape (illustrated in Fig. 3.49).

3.1.9 Textual turn-taking symbols

Textual turn-taking symbols indicate that a text or paragraph or sentence will have to assign reader a very clearly identified role as for their consistent turn taking and interpretation.

a. Major scale

This symbol, illustrated in Fig. 3.50, indicates that what follows should be interpreted as it is.

b. Minor scale

This symbol, illustrated in Fig. 3.51, indicates a modification to the marked test.

Figure 3.50. Major scale

Figure 3.51. Minor scale

c. Open or unsaturated rhythm

This symbol, illustrated in Fig. 3.52, indicates that the text is incomplete. It invites one to add more information if possible.

d. Tight or saturated rhythm

This symbol, illustrated in Fig. 3.53, indicates that the text is complete.

We may conclude that visuals presented here may all together be seen as an unambiguous, fast and effective system for reinforcing natural language because they may complement natural language with a specific set of instructions, which is consistent with the "cognitive structure of our communicative behaviour. As each stylistic choice represented reflects directly upon the syntactic structure of each sentence and paragraph, interpreting such choices systematically and attaching consistent interpretation at different levels of complexity, may significantly help as for reducing ambiguity inherent in natural language and also avoid major misinterpretation.

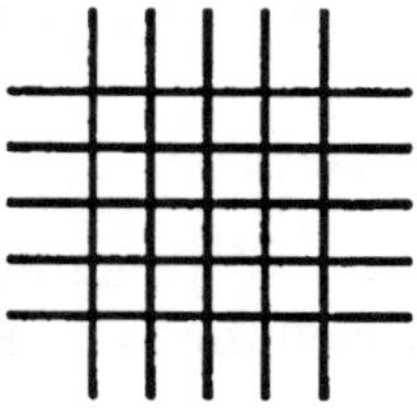

Figure 3.52. Open or unsaturated rhythm

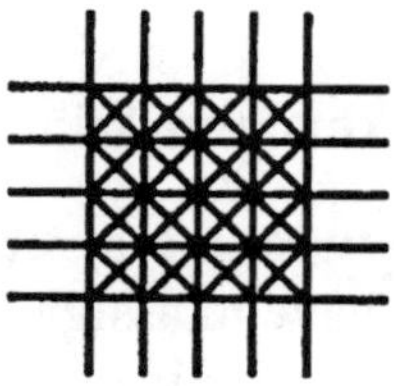

Figure 3.53. Tight or saturated rhythm

3.1.10 Conclusions

Musical notation consists of a set of technically defined terms such as andante, allegro, maestoso, etc. also conveyed throughout a set of visual symbols, carrying the composer's intention and wishes to be understood first by those performing or executing a composition. These precise indications coming in the form of words and symbols do convey what just notes written on the staff would not be able to.

When we decide to communicate by using natural language we have to think of ourselves as just some kind of music composers, because we do not just produce sentences, but we do create complex texts. And by adding specific textual instructions, which may easily complement we are practically able to monitor the communicative intercourse which any text implicitly creates. This is how competence of performance may be added to the previously identified concept of grammatical and syntactical competence each native speaker may show in a certain language. As to keep the music metaphor on, we may add that notes are syntactic categories, which form a limited set of outputs, but may combine in an unlimited number of ways.

A new framework of research will open up for linguists in order to establish the relation between text interpretative devices and the syntactical structures they reflect in different languages, by providing predictable syntactic configurations.

3.2 A new framework for describing and explaining natural language processing: a text construction environment

3.2.1 Introduction

Natural language processing needs a comprehensive and powerful metaphor that can be used to reframe complex processes as well as various phenomena in ways that may facilitate both description and explanation. Thus, a new metaphor is introduced, that can identify and name different elements playing different roles in natural language processing. In order to identify them and see them in action, a text construction environment is an appropriate and comprehensive model representing many actions that take place synchronously and harmoniously.

The model presented is not just a representation system but is meant to also support practical tools within a theoretical framework. These tools are all part of a consistent and well-defined metaphor representing the complexity of language in action, that is, communication.

3.2.2 A new way of looking at natural language

What is presented here are not just new tools but a completely new framework and way of looking at natural language phenomena. A new scientific paradigm for studying human cognition may both extend knowledge and expand an individual's potential if and only if the individual has adequate interpretative tools with which to talk about cognition, to come to know those cognitive processes that we do not see, but which we know are always active.

In the Society of the Mind theory (Minsky 1986) we see an example of what it means to create a strong, comprehensive and well-defined metaphor that describes and explains complex phenomena that would otherwise be invisible and therefore unknown.

The first need for any newly established field of discovery is to name or define the different elements that have been recognized as relevant. Indeed, no one can refer to elements that have not previously been identified and defined. It is only after having recognized and defined them that it becomes possible to refer to them consistently and to explore their relationships.

Explanation of invisible and therefore unidentified and unknown phenomena has always been necessary. This is what mythology in ancient Greece was meant to be: a description and explanation of undescribed, undefined, and therefore unexplained phenomena. "Myth is a story or belief that attempts to explain a basic truth" (Random House Dictionary 1978).

By using a metaphor, "a figure of speech in which a term or phrase is applied to something to which it is not literally applicable in order to suggest a resemblance" (Random House Dictionary 1978), a scientist may create a virtual world to explain invisible and undefined phenomena.

Virtual reality makes real those things that are unknown and may therefore be consistently and productively applied to those "unnamed phenomena" that become phenomena, "facts, occurrences or circumstances observed or observable" (Random House Dictionary 1978), only after having been experienced and recognized as such.

The representation of invisible elements or processes is obvious in disciplines such as physics and chemistry: concepts such as "force" and "gravity" in physics use visual representations and mutually shared definitions, while "transformations" or "atoms" in chemistry mean something that could not possibly be experienced directly in other ways. The impossibility of seeing and sensing such unnamed elements creates the need for highly organized, fully structured and well defined metaphors that can be used in such precise and technical ways that they become the "special technical lexicon of the field".

Unfortunately, this has not happened in all disciplines: in those that look more "real" or "realistic", researchers for quite a long time were misled into thinking that there was no need for models to represent complex and articulated processes.

Let us consider, as an example, linguistics: unfortunately, the existence of real languages and real words and the fact that individuals can actually spell and hear or write and read so-called physically perceptible texts has led researchers to think that the elements to be analyzed are merely there and available like objective data ruled by a consistent control system such as syntax. But this is not enough: language both in its syntactic articulation and lexical organization is just the surface of a much more articulated and complex set of phenomena.

Chomsky (1957) has demonstrated clearly that sentences as surface syntactic organization structures are merely the end result of a set of processes, which he named "transformations". He concluded that sentences are "generated". A dynamic view of language has been taken further to apply to meaning generation in a wider sense and to "text" (Van Dijk 1969; Ballmer 1975; De Beaugrande and Dressler 1981).

These models have all considered language as a device for the organization and representation of knowledge, this being the appropriate target to be addressed. Knowledge articulation may then be analyzed at different levels, ranging from specifically defined knowledge sub-domains to expectations about consistent actions to be taken up, to generalization and abstraction processes that can justify the interpretation process at different levels.

The belief that there is more to text than just words and sentences has also been behind theories such as the frame theory (Minsky 1975), the script theory (Schank 1977) and, later, the plan theory (Wilensky 1983) and the plot units theory (Lehnert 1981).

3.2.3 New interpretative tools

A highly articulated and dynamic metaphor is presented. It is intended to provide interpretation tools that can define and explain invisible aspects of language both within a theoretical framework and as practical tools for supporting communicative actions taking place through language. Very simply, we all agree that a word by itself is an element that may trigger a whole range of possibilities in a virtual reality of meaning.

But we may add more: we may create a virtual reality framework for both defining and explaining what is going on in our mind when we talk, write, listen or read, that is, whenever we use language to communicate, that is, to convey text carrying meaning having a specific communicative function as well as various and varying communicative intentions.

This is meant to be a much wider application of a virtual reality framework to language explanation. One thing is certain: when we talk, write, listen or read, many cognitive processes take place both synchronically and diachronically, which simply means that some of them may take place together and others will take place one at a time or one after the other.

Any model that specifically intends to justify and explain such complex "invisible reality" needs, therefore, to be a virtual reality model, based on a consistent and well-defined metaphor, because it must show the different processes taking place in different ways, at different times, at different levels and with different results.

A special meta-language intended to define, describe and explain linguistic communication, that is communication via language, must be based and grounded within a spatial and dynamic metaphorical reality with different actions taking place as well as different systems for representation. The final result will be the building of an actual text, which is the complex, highly articulated, communicative functionalized linguistic entity in the process of progressive structuring, also defined as construction. A text construction environment needs to be a strong metaphor representing the various cognitive activities that take place, just as different tools are needed to build up a text.

A virtual reality environment for text construction therefore allows the interactive user to be primarily a text builder who experiences text building directly. From now on users will be called choosers as, by assuming such perspective, they will interact responsibly with the environment, not merely trying the different possibilities but directing and acting on the environment itself to ensure the result corresponds to their intended plan.

Let us hope that the virtual reality community will agree with my decision to redefine the user as the chooser because this is what virtual reality should actually be for: amplifying and giving multiple possibilities, none of which were visible or in existence previously. 'Choosing' means navigating in a wider space of possibilities, interacting with the various scenarios that may occur and then making a decision based on visible criteria.

Controlling a certain process means being able to experience it in its phases by amplifying one's own perception of it. The discovery of a new territory and the exploration of invisible and therefore unknown and un-represented elements happens many times in science; for example, we have already considered chemical transformations and physical models for knowledge representation.

What we see in such cases is the use of metaphors to represent something that needs to be described, defined and explained, before it may become fully known and used by a research community and a wider audience also. Once a metaphor is understood, it may be used to refer to specific phenomena.

In order to talk about different processes, Minsky (1986) has developed metaphors such as "society", "agency" or coined new names and definitions such as "polinemes", to provide terms to define and describe activities that were invisible.

3.2.4 The "nature" of CPP-TRS

CPP-TRS (Tonfoni 1989-94) is both a methodology and a language. Through the methodology (CPP) those invisible aspects of communication are identified and represented by the meta-language (TRS), which complements natural language. Icons are not intended to represent words or sentences as a substitute for them; they are rather intended to represent visually what natural language does not convey naturally.

CPP-TRS is a visual language based on a system of 12 canvases, 10 signals, 14 symbols and a set of other dynamic visual schemes. It is based on the fact that every communicative action is the result of a set of cognitive processes, and the whole system is based on the concept that communication may be enhanced by visual planning and organization, therefore raising the visual perception of text. Based on a very simple syntax, it is capable of representing meaning and intention as well as communicating function visually. CPP-TRS is an unambiguous, fast and effective system for reinforcing natural language. It complements natural language by adding certain important elements that are not represented by natural language itself.

A certain set of elements, which we do not see because they are not represented by natural language, such as cognitive styles, are identified and named with a verb, with a technical meaning and a corresponding and consistent symbol; communicative types are represented with a name, with a very specific definition and meaning and a corresponding signal.

What we have here is a qualitative model for communication. Once a basic understanding of cognitive styles and communicative types has been reached, the qualitative model of text will turn the quantitative aspect, that is, sentences and paragraphs, into a qualitative representation, in other words, communicative intention and function expectations. However, symbols and signs are just one aspect of it: there are several other visual dynamic schemes.

Textual machines represent physically and dynamically those cognitive processes that go on in the mind from the most simple to the most complex. They are models that explain things that cannot be seen, a process with different stages that evolve dynamically in time.

The model based on canvases, symbols, signs, textual objects and textual machines can be defined as a system that represents visibly that which is invisible in language and that physically models cognitive processes in action. By providing the chooser with symbols, signs and machines, it is also a toolkit. CPP-TRS is therefore both a descriptive, explanatory model dealing with the complexity of text understanding, and an active system for text generation, based on a new way of conceiving text production and text processing. By understanding the conventional meaning assigned to basic signs - the geometric shapes of squares, triangles, etc. - the chooser will appreciate the meaning of objects that are built up by combining the basic shapes.

To give a very simple example: if a square represents a narrative text, then a cube, which is made of six square elements – faces -, is a 3D representation of narrative text, where each face may be a chapter of a story or a short story in itself. The cube cannot represent another type of text because it would not be consistent, and would be misleading in that it did not respect the agreed semantics.

Textual machines also have this consistency of shape as they are comprised of the same basic shapes that carry specific meanings. The chooser, by learning the basic meanings, can thus learn easily the entire system.

This model is complex in the sense that it deals with many different aspects of communication, given that communication and language are themselves complex. Complexity does not imply difficulty; rather a model that represents a complex phenomenon needs to be simple. If it is not simple, it does not help to describe, define or explain the fact or set of facts that it is intended to represent. CPP-TRS provides a specific, agreed terminology that uses technical terms in a very precise way, in order to refer and identify the different aspects of a complex process and name them.

3.2.5 A text construction environment

A text construction environment is a virtual environment intended to represent the process of planning and structuring of a text. It is based on a selection of visual models and metaphors that represent different actions taken in time, both synchronically and diachronically, to illustrate "text construction". The term "text construction" is used instead of text production or text generation because by construction we can refer to the material to be processed, which is information, to the different tools performing different tasks in time, which are cognitive processes, and to the different shapes and packaging modes the processed material, which is knowledge, may acquire to become a text.

The final result of such operations is the construction of a text. Let us now have a look at Fig. 3.54.

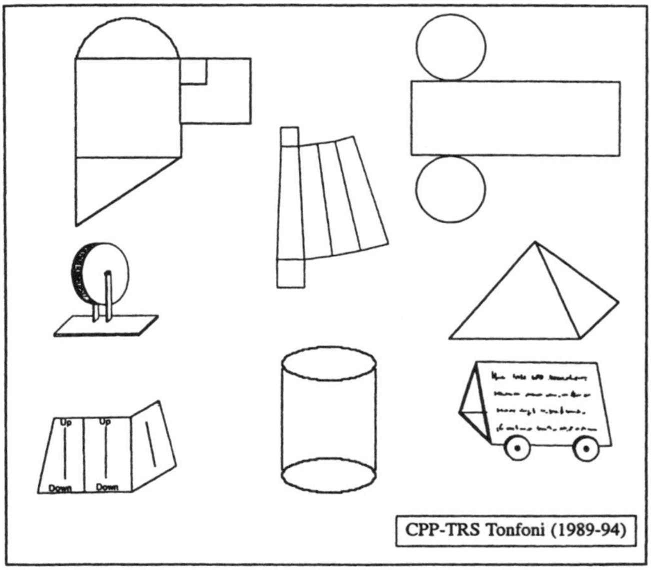

Figure 3.54. The metaphorical construction environment for building a text

The metaphorical construction environment in Fig. 3.54 shows the progressive building up of a text. It is based on a whole set of shapes and indicates unambiguously and directly the different communicative functions that the specific text construction will entail. These functions are part of the foundation to ensure that from the planning stage they will remain in the actual building and appear in the final text.

Around the foundation is a set of textual machines that demonstrate the many and various cognitive processes that will take place at different times to turn unprocessed information into sentences, paragraphs and text units, representing processed and packaged knowledge.

Fig. 3.54 shows some textual objects that can be added to the text construction to represent other communicative actions, derived from the text according to the specific function each object has been assigned within the CPP-TRS framework.

We saw earlier the textual signs that indicate the various aspects of text development. Textual shapes may be further developed and may evolve into textual objects that show consistency and coherence between the specific shape they have and the kinds of text they contain.

Each textual object carries with it evidence of its particular meaning. Just as the shapes of traffic signs carry specific meaning that is immediately understood, when text is shaped to form a particular textual object, the text shape itself conveys the kind of information needed to interpret text consistently.

In CPP-TRS six textual objects are presented:

a. Cube

A cube, illustrated in Fig. 3.55, is the individual and collective text-shape and space organization for constructing a story. To construct a story, text constructors think of the text of the story as a cube consisting of six flat, square facets. Each square facet represents a segment of narrative text that tells part of the story. Text- constructors begin by constructing the text for each square. To create the story, they fold the six textual squares into a cube. The arrangement of each textual square on the cube demonstrates how the parts of the story relate to one another.

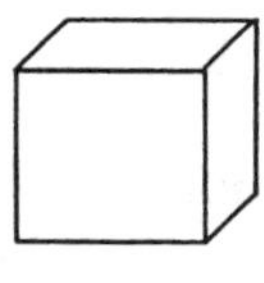

Figure 3.55. Cube

b. Pyramid

A pyramid, illustrated in Fig. 3.56, provides individual and collective text-shape and space organization for a story. Text constructors present a story that reminds them of other stories, which they add to the initial one. The same story may remind them of other analogous stories as well, which may then be added too. The pyramid is different from a cube in that it contains personal memories related to a story rather than the story itself.

Figure 3.56. Pyramid

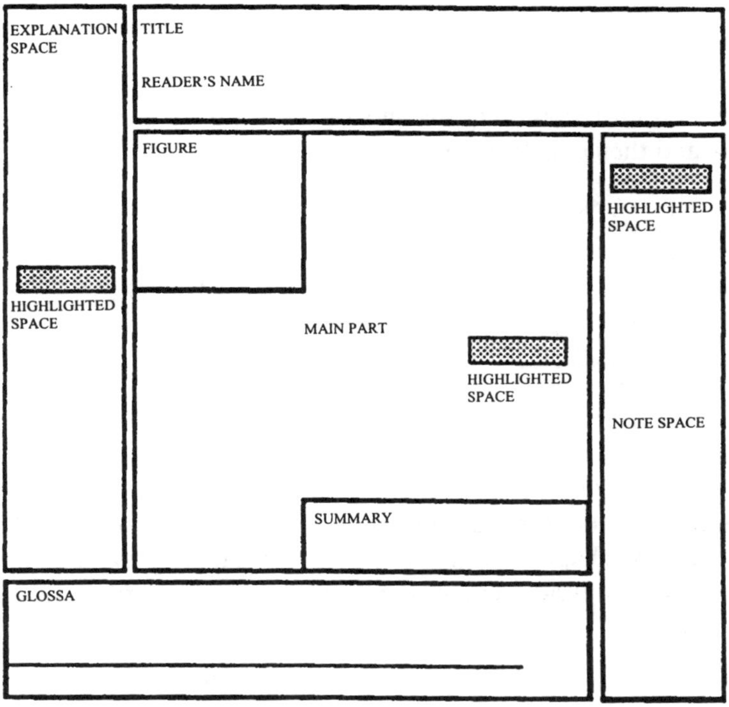

Figure 3.57. Miniature

c. *Miniature*

A miniature, illustrated in Fig. 3.57, provides individual and collective text-shape and space organization for expanding and completing texts. Miniatures were a form of ornamental writing used in the Middle Ages to make the perception of certain texts more pleasant. A miniature textual object has definite parts each containing different kinds of information.

d. Polittico

A polittico, illustrated in Fig. 3.58, is a text-space organization for raising and discussing a topic. Historically, a polittico is a picture made of hinged planes that fold like doors in front of the main scene. In polittico text constructors raise a topic and ask for comments and opinions.

e. Obelisk

An obelisk, illustrated in Fig. 3.59, is a collective text-shape and space organization for recalling a certain story or set of stories well known to a group, and then constructing a common story.

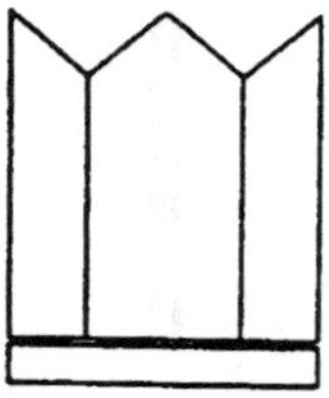

Figure 3.58. Polittico

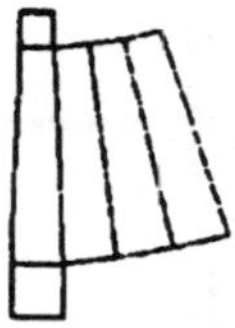

Figure 3.59. Obelisk

f. Cylinder/column/papyrus

A cylinder/column/papyrus, illustrated in Fig. 3.60, is an individual and collective text-shape and space organization for explanatory text construction. A cylinder/column/ papyrus is a combination of a cylinder or column and a papyrus, allowing text construction on both the outside and the inside.

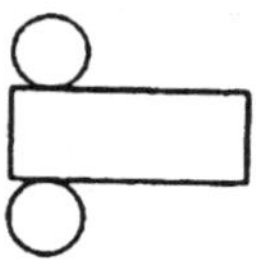

Figure 3.60. Cylinder/column/papyrus

3.2.6 Conclusions

The whole metaphor of a text construction environment is a model for the identification and naming of aspects of natural language that cannot be seen and are therefore not usually taken into account. This metaphor provides words that name these aspects and make them obvious. When they are experienced through imagining them in action, they become suitable tools for both self-reflection and further learning.

3.3 Text Processing as text composition and text execution

3.3.1 Introduction

This work provides a new framework for describing and explaining communicative actions involved in text organization and text understanding. Based on a well defined metaphor required by the complexity of the task itself, the model provides specific visual tools that help us recognize, identify and name those hidden aspects of natural language that are so important to text processing and communicative intercourse.

CPP-TRS (Tonfoni 1989-94) is both a methodology and a language. Through the methodology (CPP) those invisible aspects of communication are identified and represented by the meta-language (TRS), which complements natural language. CPP-TRS is a new paradigm where icons or visual schemes are not substitutes for words or sentences, but rather represent visually what natural language does not explicitly convey.

Current trends in text processing were illustrated in a recent article in AICOM, vol. 7, no. 3/4, by S. Geldof, a report on the 1993 Bolzano International School.

Referring to the different approaches that were described, Hovy and Nirenburg (1992) defined "an interlingua as a notation used in Machine Translation systems to represent the propositional and pragmatic meaning of input text" and proposed an organized structure indicating text performance.

On the other hand Kay (1986) underlined the point that bridging the gap between text and interpretation involves "guessing" and that the success of interpretation depends largely on the "ability of the hearer to interpret hints".

Linguists and artificial intelligence researchers concerned with natural language processing must always take into account the complexity of natural language. Solutions to problems arising from this may be different but complexity always needs to be dealt with.

3.3.2 How CPP-TRS can contribute to managing complexity in natural language

CPP-TRS is a visual language based on a system of 12 canvases, 10 signs and 14 symbols and a wide set of visual dynamic schemes. CPP-TRS is an unambiguous, fast and effective system for reinforcing natural language, which complements natural language by adding certain important elements that are not represented by natural language itself, such as the communicative function of text and the communicative intention of each paragraph or sentence, as well as the role the text receiver is supposed to play.

Musical notation uses a set of visual symbols to convey the composer's intentions and wishes to those performing or executing a composition. These symbols communicate what the notes written on the staff alone cannot.

Text producers are therefore called text composers because they will not merely produce sentences, paragraphs and text but will also provide the textual instructions that complement text itself as text execution cues.

Text receivers are consequently called text executors because they will not merely receive sentences, paragraphs and text but will also be provided with the visual organization of text instructions for interpreting and understanding the means for executing text.

Text is conceived as a musical composition: the text executor "plays" a text, just as a musician executes a composition. To indicate how a text is to be executed, 10 signs and 14 symbols are provided.

The text composer is invited to start from a global perspective of communicative function of the text or the text units and move to the intended meaning of the single paragraph and sentence which is subject to change. The text composer may also ask the text executor to take some specific action by conveying visually some specific requests.

As in musical composition, by using notations a text may be controlled both on a global and local level to facilitate the interpretation process, since CPP-TRS is a consensual system for representing invisible aspects of natural language visually and turning them into specific instructions for text execution.

If text processing may be seen as both text composition and execution, it is evident that the starting point is never merely the sentences or paragraphs but rather the visual organization of the text composition, where sentences and paragraphs are the final result of the process.

According to such a perspective, there is no such process as generic sentence organization where words and sentences flow automatically and unambiguously convey a specific meaning, since this is never really true in actual communication.

3.3.3 The new approach "from global to local"

Instead of starting with words and stringing sentences together, text composers will have to decide the communicative function to be attributed to text, the communicative intentions that will vary in the paragraphs and sentences, and the role they will ask the text executor to play. They will then design the overall communicative structure of the text to go with the text itself. The CPP-TRS-based global and local intention indicators are presented in Fig. 3.61. Starting from global control indicators, we first find signs, global control icons that indicate communicative functions within the text. Moving to local control indicators, we find two kinds of symbols.

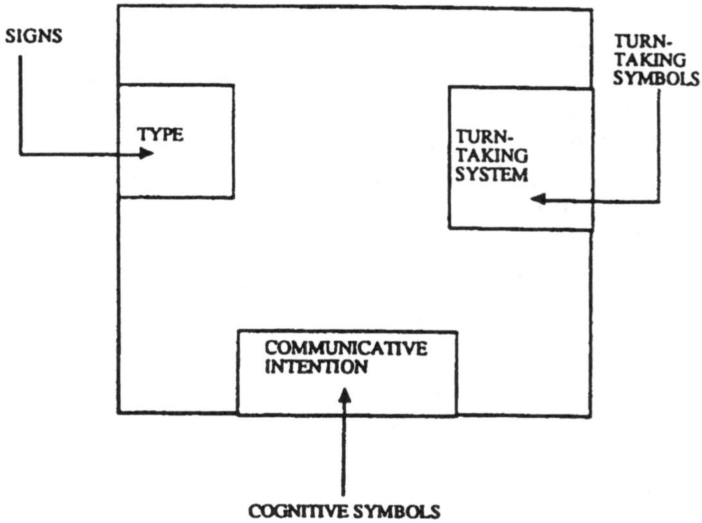

Figure 3.61. Global and local indicators

The first, called cognitive symbols, indicate, paragraph by paragraph or sentence by sentence, communicative intentions that represent specific cognitive processes. Each of these symbols has a technical meaning that relates to a cognitive process and identifies a specific intention to be indicated within the text composition for consistent text execution.

The second kind of symbols, called turn-taking symbols, indicates the kind of interaction the text composer wants to have with the text executor. Turn-taking symbols are meant to indicate when, how and why the text executor should actively become part of the text composition during the text execution process. Turn-taking symbols are the most interactive indicators, allowing the text executor to change, add or remove parts, which are in the text composition.

3.3.4 A mental keyboard to compose and play texts

The composing of a text implies a previous organization of text both on a global level – signs - and on a local level – symbols -. CPP-TRS may thus be defined in terms of a visual notation system that is as consistent as the system of punctuation.

The system of punctuation is conventional in that it was created *ex novo* and consensual in that everybody agrees to use it in a particular way. It does not change and represents pauses and intonation in language. The punctuation system complements natural written language and adds to it what is there in oral language but would otherwise be missing in written texts.

The present system is not different to the punctuation system in that it represents visually those elements, such as communicative intention, communicative function and communicative turn-taking, that are invisible but important to the understanding of a text in the way the text composer intends. CPP-TRS is both a conventional and natural meta-language that makes explicit from the beginning what otherwise is left to arbitrary interpretation. Making text more explicit by adding visual conventions enhances the final understanding by providing the appropriate "hints of guessing" (Kay 1986).

Let us now look at Fig. 3.62, which shows the CPP-TRS-based keyboard.

In a CPP-TRS (Tonfoni 1989-94) perspective the traditional process of creating a message is reversed. Instead of starting with words and sentences, text composers must first understand their own intention, then design the structure of the text, and finally select sentences and words for it. This approach avoids the limitations of the traditional approach by making communicative intentions explicit at the very beginning.

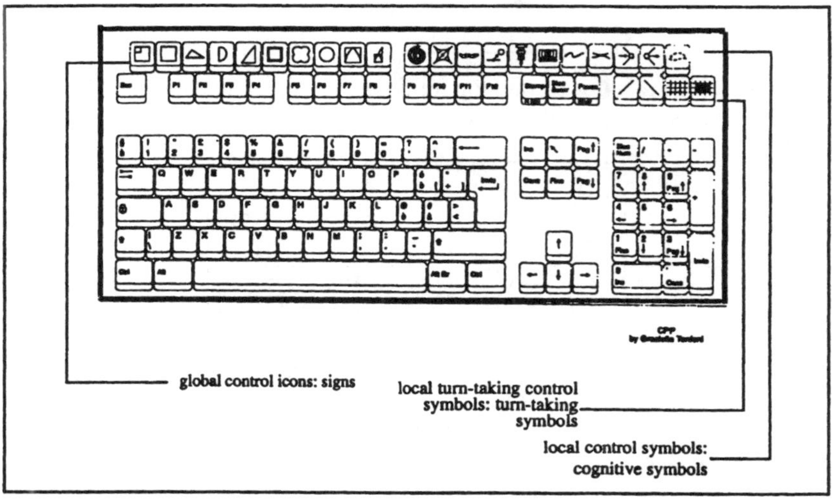

Figure 3.62. The CPP-TRS-based keyboard

3.3.5 A mental scanner to segment, mark off and execute texts

The executing of a text by interpreting notations allows text segments that must be played according to specific instructions to be identified. Just as a scanner is able to reproduce and reprint a certain text, a mental scanner metaphor helps natural language experts to program the text for a certain operation that may replicate or modify the original text structure, either locally or globally. By keeping notations transparent, it will always be possible to trace back the kind of operation performed on the structure of the text.

The CPP-TRS based scanner is shown in Fig. 3.63, in Fig. 3.64 and in Fig. 3.65. The transparent use of it allows the text executor to read signs and symbols both with the text and as a visual language on its own, that carries information about text that would otherwise be left implicit and not linguistically conveyed.

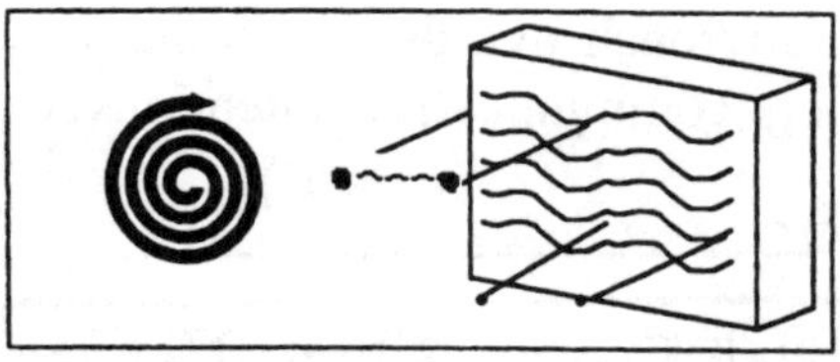

Figure 3.63. Stage A – Local scanning for communicative. intentions by identifying segments and marked-off portions of text

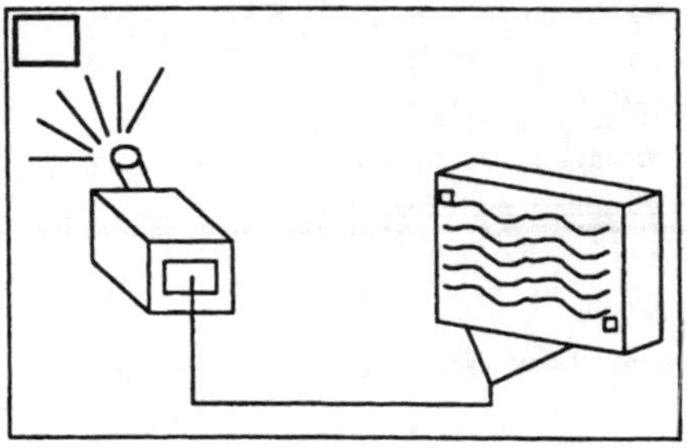

Figure 3.64. Stage B – Global scanning for searching and identifying communicative functions of the text or text unit

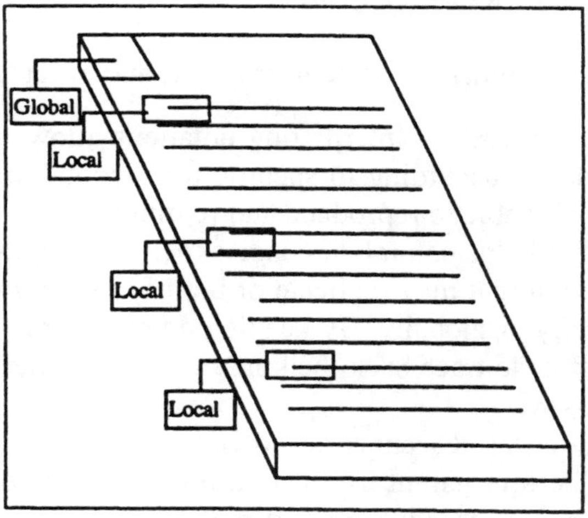

Figure 3.65. Stage C – Globally and locally scanned text or text unit ready to be executed or reformulated either globally or locally according to turn-taking instructions

3.3.6 A toolkit for text composition and execution

In order to provide natural language research with both a new terminology and new metaphors to indicate elements and aspects of language that are otherwise vague, subject to personal interpretation or merely undefined, new definitions of specific cognitive processes and new icons for representing invisible aspects of language have been introduced, described and explained.

CPP-TRS is a unifying framework that allows research in natural language processing to take different directions, to be supported by a commonly shared terminology while still retaining its originality and diversity of perspectives.

The variety of approaches that characterize natural language processing research may be considered a sign of a positive attitude towards the many ways of managing the complexity inherent in natural language. A common framework may help better to identify common problems and different ways to approach them.

The system may thus be used for very practical purposes in various domains. The mental keyboard, presented above, is a metaphor for rethinking the text composition process, but it may also be turned into a prototype keyboard, with signs and symbols available for transparent use in text processing as an extension of the punctuation system and word processing facilities.

The same is true for the mental scanner, which is on the one hand a metaphor for rethinking text execution through a physical and defined model, but may also be considered a model of an actual scanner, where signs and symbols are made available for filtering, filing and revising processed texts.

3.3.7 Text compression

The CPP-TRS based theory of text comprehension provides an accurate and articulated framework for attributing qualitative interpretation to a text globally or to text segments locally and is therefore a useful model for representing text execution at both levels. We will now describe the theory of text compression.

"To compress" means literally "to press together and force into less space"; a compressed text is a text in which information density is being reorganized not just quantitatively but qualitatively with a multilayered distribution and in a different textual space.

The difference between a quantitative and a qualitative approach to information density in text and text segments can only be conceived within a quantum theory of language based on qualitative reasoning about communication.

CPP-TRS is a model for text execution and, according to a set of interpretative cues, it is a qualitative model that generates a model of text that is ready to be interpreted and executed by explicating visually qualitative information about each text unit, as can be seen in the two examples of Fig. 3.66 and Fig. 3.67.

The text unit means a further step in text processing, where not only local and global control interpretive cues have been identified as in text and text segments annotation, but where an adjustment has also taken place involving the temperature-pressure relationship and balancing. A text unit is the first step for inorganic text production and is aimed at reducing the natural distance between text producer and text perceiver through the use of explicit interpretation cues. The theory of text compression aims toward reducing the distance by significantly shortening the existing bandwidth in current interaction and by keeping the interpretation cues transparent.

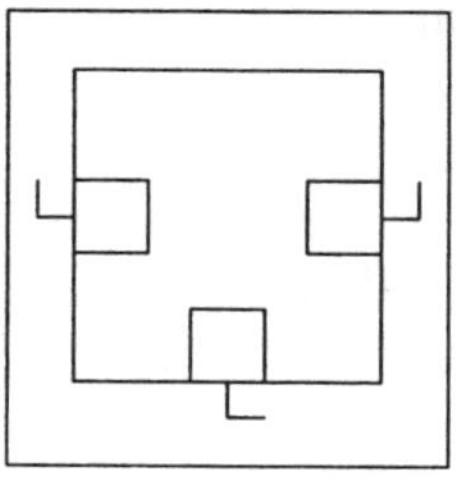

Figure 3.66. Example of visually quantitative information about a text unit

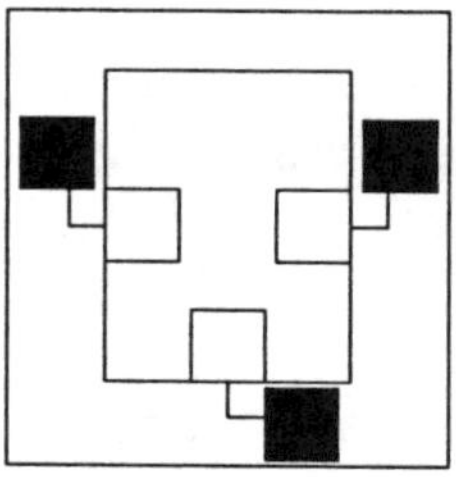

Figure 3.67. Another example of visually quantitative information about a text unit

Therefore, a CPP-TRS based model for text compression does not operate directly from texts, rather from a processed text that shows qualitative information about text execution through a consistent application of signs and symbols. Signs controlling text globally do not have to undergo any further change during text compression. Symbols controlling text locally will activate a further process, which are named multilayered distribution of text segments.

In order to understand what this process consists of let us now have a look at Fig. 3.68, which represents a high compression device for text segments multilayered distribution.

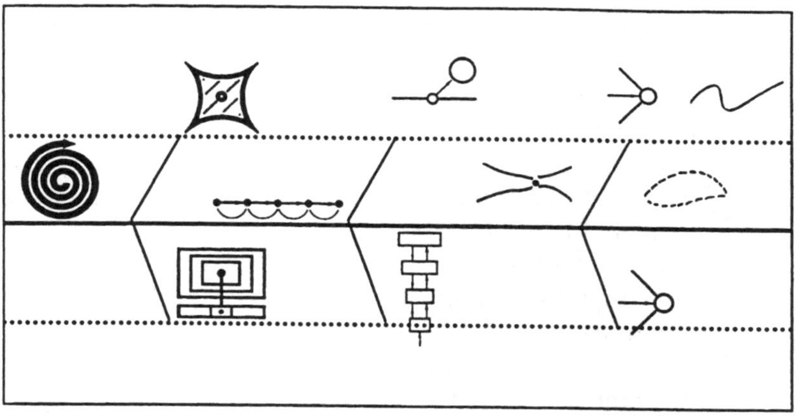

Figure 3.68. High compression device for text segments multilayered distribution

As a consequence of three-layered distribution each text is configurated according to a dynamic process, which will visually and consistently represent the quality of information. A processed text which undergoes a process of text compression, will look like the illustration of Fig. 3.69.

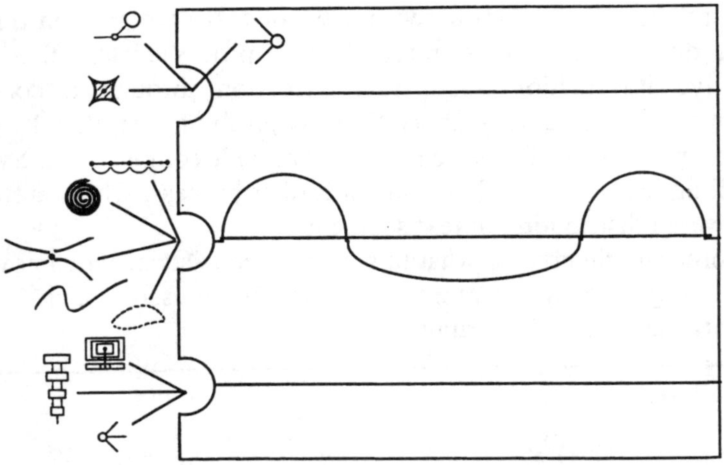

Figure 3.69. Processed text undergoing a process of text compression

The three-layered distribution of text segments according to the local control symbol is in fact a reduction of a ten-layered distribution. The upper level corresponds to those local control symbols representing a lower density of information; the intermediate level corresponds to those local control symbols that represent a medium density of information; and the lower level corresponds to those local control symbols which represent a high level density of information.

By checking the measure of the circles, the exact symbol may be detected by the use of a textual template. Qualitative compression is the result of a very accurate process of information packaging based on inorganic textual configurations, which may be defined as CPTRU, Communicatively Positioned Textually Represented Unit.

Configurations are topic-driven and constitute information territories, which are qualitatively and quantitatively different from the text from which they are derived. The process each CPTRU has undergone is visible both at the level of global control signs through the use of a textual mask and at the level of local control symbols through the use of a textual template.

The access to global information is the first step in information territories visualization, whereas access to local information is regulated by a further level of visualization. Once information is compressed, which means that it has undergone major structural changes, the CPP-TRS framework allows for full visibility of the processes. In other words, any text and text segment undergoing a compression process will keep visible traces of what has happened to it.

Any decision to promote transparency in communication has been maintained at every step, including the planning of a transparent mechanism both for quantitative and qualitative text compression. It should also be pointed out here that text compression applies to natural language. Such processes cannot be transported to either visual or gestural states of information, since they radically differ in their properties and characterization.

3.3.8 α, β, γ radioactive information territories

A processed text unit that acquires a high level of communicative visibility becomes an inorganic structure aimed toward increasing effectiveness of communicative processes and intercourses by reducing the bandwidth between text producer and text perceiver.

A compressed textual unit is therefore an extremely powerful unit both for information and communication visibility, and its structure is indexed, stored and retrieved, based on a topic or keyword identification. A whole set of topic driven compressed textual units constitute information territories.

Accurate exploration of information territories is very important in order to detect the different amount of qualitative information that is provided about a certain topic. Inorganic textual structures may undergo further processes, which we will now consider. Transformational processes that occur on a compressed unit may modify the communicative energy value and the information force value as well at different levels, and they may generate radical changes as well.

This is the reason to postulate a kernel theory of communicative energy transformation and identify three different kinds of radiations that may be active, each modifying it more or less radically. These are named as α, β, γ radiation in close analogy with the radiation found, analyzed and reproduced in nuclear energy phenomena. These kinds of radiation apply to compressed structures at different levels and produce qualitatively different kinds of reactions. Just as in nuclear physics, they perform on a qualitatively different level. Let us now analyze them in details.

A radiation applies first to topic-driven CPTRU territories and then to sign-controlled CPTRU. When applied to a topic, α radiation will enlighten, that is, activate a certain CPTRU topic-driven territory and will aggregate according to the communicative patterns configuration assigned.

In the following Fig. 3.70, an example of α active aggregation is illustrated.

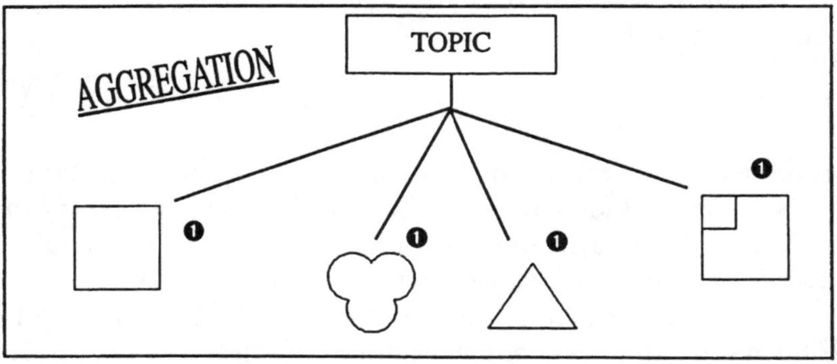

Figure 3.70. α active on CPTRU topic-driven territory

When applied to a sign, α radiation will then enlighten, that is, activate a certain sign-controlled CPTRU, and will attract according to the communicative patterns configuration assigned.

In the following Fig. 3.71, an example of α active attraction is illustrated.

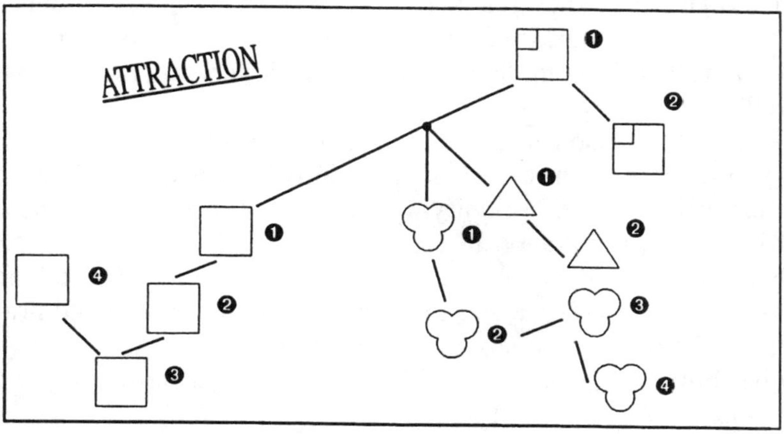

Figure 3.71. α active on sign-controlled CPTRU

α radiation produces acceleration, but cannot create transformation because both topic and sign represent a stable component that does not allow any change at the core that may affect the CPTRU structure. Nevertheless α is able to activate, aggregate and attract; thus it can be defined as that kind of radiation that may produce acceleration but no transformation; α active CPTRU structures may aggregate and attract.

γ radiation applies to the kernel of a CPTRU. It can activate and excite a specific locally controlled textual segment or set of segments that is symbol-controlled. Thus it is able to trigger a reaction. Such a result may be defined as a transformation because it may completely alter the qualitative nature of the symbol-controlled textual segment or segments. Both new and transformed segments must keep a visible trace of the reactions they have been undergoing.

In the following Fig. 3.72, an example of γ active excitation is illustrated.

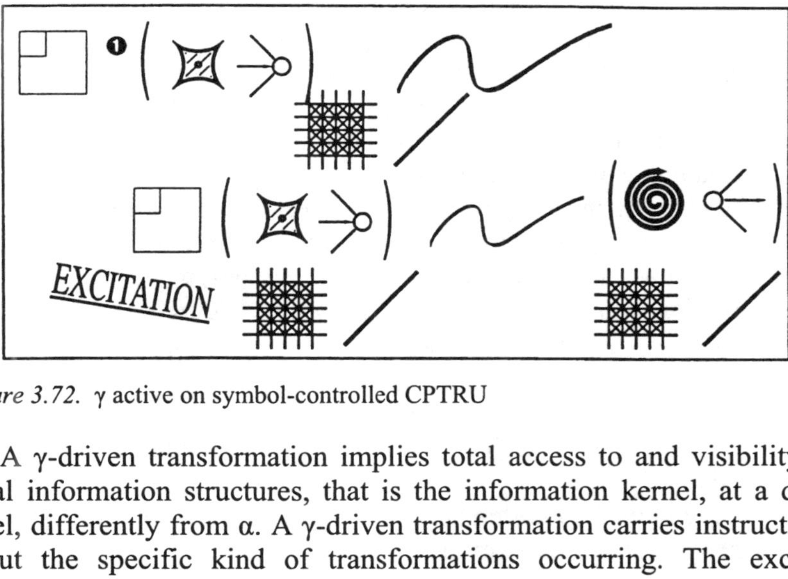

Figure 3.72. γ active on symbol-controlled CPTRU

A γ-driven transformation implies total access to and visibility of local information structures, that is the information kernel, at a deep level, differently from α. A γ-driven transformation carries instructions about the specific kind of transformations occurring. The excited CPTRU will therefore retain its original state as well as the transformed one, leaving a trace. To conclude, γ radiation is able to produce excitation and transformation by leaving a trace but no acceleration. It applies to symbol-controlled text segments, that represent the unstable component.

When γ active CPTRU is excited and keeps a trace, α radiation will necessarily be active as well. β radiation has been left to the last as it is the most powerful and radical of the three kinds of radiation. B radiation applies to the kernel of a CPTRU with consequences and side-effects for the overall system; it activates, excites and explodes many or all locally controlled text segments that are symbol-driven. It is therefore able to trigger a very powerful reaction, the strongest one among those we have considered. This is a transformation because it completely alters the qualitative nature of the identified symbol driven-text segment or segments.

As in earlier examples, both new and transformed text segment or segments will retain a trace of the reaction. An active β implies that α and γ are active as well and it causes very strong side effects. If β applies to larger text segments it may alter the qualitative nature of the whole CPTRU to a point where even the more stable component, which is sign-driven, may actually disintegrate. A high level of control therefore needs to be kept on the changes made.

If γ radiation is very strong, it may produce a high level explosion causing α and β to be active to a point where they produce both transformation and acceleration. In such case acceleration is produced as a consequence of a strong α activation to cope with the radiation of γ activated by β. As in γ, any β-driven transformation implies total access and visibility of local information structures at a deep level.

β radiation is able to produce excitation, explosion and transformation by leaving a trace and may also trigger α into producing acceleration as well. This applies to turn-taking symbol-controlled text segments, which are linked to local control symbols and together represent the unstable component of CPTRU.

In the following Fig. 3.73, an example of β active excitation is illustrated.

To conclude, β radiation always triggers transformation and may sometimes produce acceleration as well.

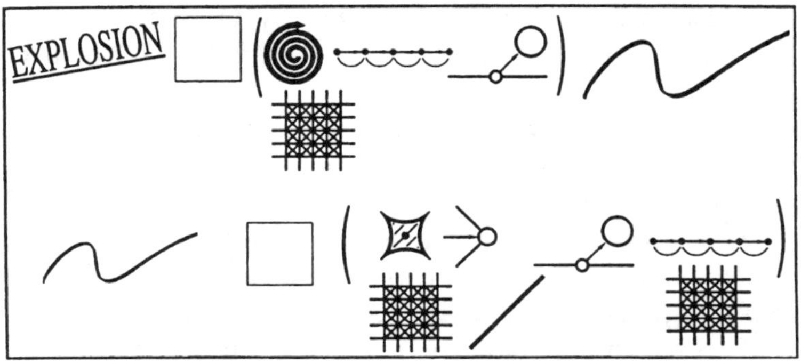

Figure 3.73. β active on turn-taking symbol-controlled CPTRU

A β active CPTRU, while exploding, will activate γ and α as well as retain a trace. The kind of described phenomena are the result of operations that need to be controlled each time. They can be of tremendous help in raising transparency in communication. This is just like X-raying texts to facilitate dramatically information exchange based on a shared system for interpretation.

3.3.9 Conclusions

A metaphor that represents a global framework for modelling a well-defined and complex research domain, such as natural language processing, may become a whole set of tools.

This is what the currently described model and meta-language are intended to do by defining the "context of occurrence" of each given text. We may also see the overall system as a "natural sublanguage" used in reference to a whole variety of restricted domains.

3.4 On visual text planning, processing and programming

3.4.1 Introduction

Relevant research is taking place in the field of the integration of natural language and vision processes. Significant analogies can be found between visual perception and language perception, as pointed out by Winston (1992). The weakness of interactions between image processing and natural language processing was found by Marr (1982), Ballard and Brown (1982) and Joshi (1991). A common semantics of language and vision has been thoroughly analyzed by O Nuallàin and Smith (1994). An important exploration in the field is by McKevitt (1994a, 1994b).

An interesting survey on literature of computational models for integrating linguistic and visual information is by Srihari (1994). Research on iconic languages by Mealing and Yazdani (1990) has shown the need for accurate research in icon production.

Research into the integration of natural language and vision has grown very large, as we can see in McKevitt (1994c, 1994d).

CPP-TRS by Tonfoni (1989-94) can be defined as a computational model for integrating qualitative reasoning about natural language texts and visual representation of the newly defined qualitative communication categories.

3.4.2 Visual text planning

The basic metaphors here introduced are taken from drawing and painting. The first metaphor is the canvas. For visual artists a canvas is a basic support for their work and is the first consideration for execution of the work. A canvas can be fine and smooth or coarse and rough, depending on the tightness or looseness of its weave. A visual artist chooses a canvas partly because of the particular characteristics of its texture. A finely textured canvas is appropriate for detailed, precisely drawn or painted work, while a coarse canvas may be worked for a thick, bold application of paint. Thus texture (Tonfoni 1989-94) has a radically different meaning from that found in current literature.

Coherence of the text is investigated in terms of texture, a lexico-grammatical pattern that signals the relationship between individual messages and is identified in terms of cohesive ties and cohesive devices (Halliday and Hasan 1985). Coherence is therefore investigated with respect to register in terms of text-structure as related to a generic structure potential (GSP) associated to text (Halliday and Hasan 1985). Coherence as text structure is determined by semantic relations that describe how sentences and text units are linked together to form and continue discourse (Hobbs 1979; Mann and Thompson 1987).

Texture in CPP-TRS is a set of unprocessed sentences and text paragraphs ready to be planned and organized according to the communicative intention of the text planner. Just as a visual artist chooses a canvas with an appropriate texture, a text planner needs to decide which kind of "texture" a text should have. The texture of text has to do with the density of information conveyed (quantitative level). Depending on the particular communicative purpose of the text, what the text planner intends to achieve and what the text is actually for (qualitative aspect), the texture of the text may be more or less dense. This means that the qualitative aspect of language will regulate and control the quantitative level.

Not only does the metaphor of canvas carry the property of texture, it also provides a visual context for the process of text planning. After planning a canvas with the right kind of texture, that is a consistent quantity of information, the planner establishes perspective, point of view and focus, and possibly a cliché for the drawing or painting. The perspective identifies how text planners position themselves; the point of view shows what the text planner selects as most relevant; the focus indicates what is topical; and the cliché displays how the text planner deals with organizing a communicative space in which the reader may move around.

According to this perspective, the reader is the visual text processor. A visual text processor is an active individual who interprets the visual text planner's communicative intention, as well as the communicative function of the text, that is the qualitative level of communication within the text itself as articulated in sentences, paragraphs and text units, representing the quantitative aspect of information.

Text processing may therefore be enhanced by bringing to light those aspects of meaning that are invisible, because they are not conveyed linguistically within the text but are relevant for text interpretation. To cope with this problem, both global and local planning of text is conceived visually. Only after texture is organized visually may the text planner use sentences and paragraphs.

This approach is consistent with the Society of Mind Theory (Minsky 1986), which contends that many specific cognitive processes take place before actual language, sentences and texts are formulated. Difficulties in text processing are therefore explored and addressed before text occurs, in the cognitive planning of it. Difficulties we encounter in organizing a text are usually the result of cognitive, not linguistic, problems.

Text planners are asked to start from a global perspective on the communicative function of the text they are going to organize and to move towards a local consideration of the communicative intention of each text unit, paragraph and sentence. They must also define the role they want the text processor to play.

CPP-TRS provides a set of canvases that are visual schemes and global and dynamic representations of communicative actions that may be taken in the course of text planning. Canvases are thus visual dynamic schemes that represent the various communicative actions that may be taken and are therefore self-reflective tools. Once text planners have an overall global view of their textual plan by visually representing it, they may then proceed to a detailed organization and structuring of the text itself.

Before organizing a text, text must be conceived as an open possibility, which can be further planned from a multitude of perspectives and points of view, with many different focuses, intentions and according to many organizational considerations.

Canvases visually represent possible actions that could be taken to plan text and they retain the visual planning of the linguistic part, that will emerge in the final result. Let us now look at the overall model, which both identifies and names the different processes and actions that take place when a text is structured. It provides the text planner and the text processor with new visual tools that can make visible what is invisible within the text itself.

By enhancing visual text perception, both text planner and text processor share visual semantics about the qualitative level of the text. This approach therefore differs radically from traditional approaches to organizing text, which assume that individuals create meaning and intentions after having connected words and sentences. In fact, planning of meaning and intention comes first, with words and sentences coming last, as we can see in Fig. 3.74.

Canvases are visual dynamic representations of actions, which need to be used in text planning on a very general level. From a canvas, text planners go directly to linear text organization, keeping the visual schemes together with the text itself. They may be defined as visual text planners because they explicitly declare the communicative plan they have chosen by making it obvious and keeping it with the text.

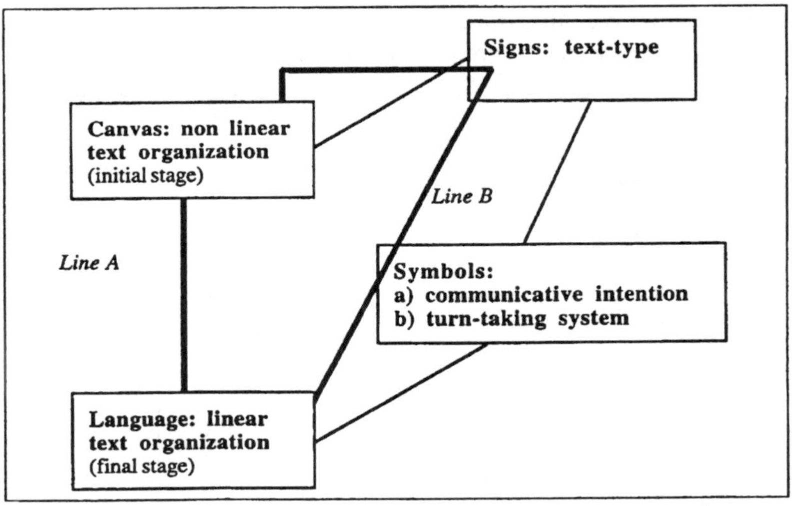

Figure 3.74. Role of canvas, signs and symbols in planning and processing text

As a result, the text processor will be able to interpret the text more easily by knowing the communicative plan and by being directly assisted in the interpretation process being told the kind of communicative action that has been taken. The visual processor will therefore receive both quantitative information, which is sentences, paragraphs, text units, and qualitative information, which is the kind of communicative action taken to produce the text.

By proceeding from a canvas into linear text organization the visual text planner keeps the overall general plan of the text and the kind of action taken (Fig. 3.74, Line A). By "taking" Line A visual text processors receive a general level of qualitative information, which is much more than what they would receive from a traditionally processed text without the qualitative information.

But the visual text planner may wish even more qualitative information to be added to the text and to be available to the visual text processor by accessing other visual tools which are provided.

Let us now follow Line B in Fig. 3.74. Once the visual text planners have given a general view of the kind of action taken on text, they may then move towards the more detailed planning of different paragraphs and sentences both on a global and a local level. The system starts from a global perspective of communicative functions, and moves towards the role the text processor should play, as shown in Fig. 3.75.

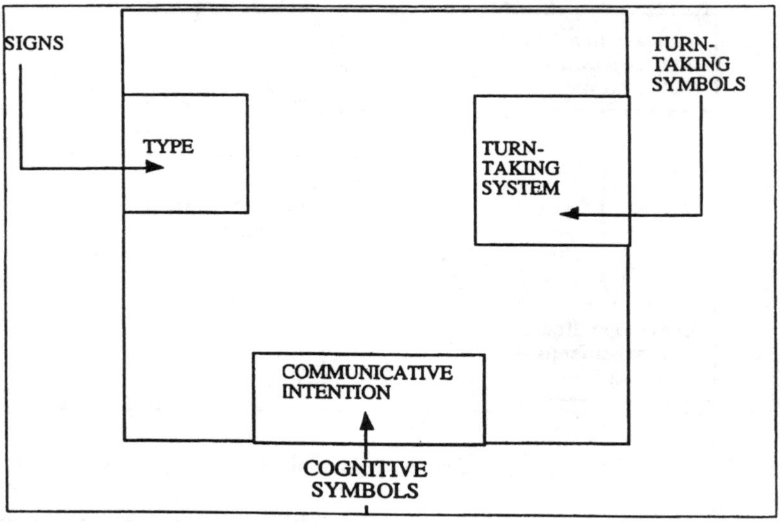

Figure 3.75. Role of signs and symbols in processing text

Signs are global control icons for explaining the communicative type and function of a text unit. Symbols are local control icons for explaining the communicative intention and style of sentences or paragraphs.

Turn-taking symbols are local control icons for explaining the role the visual text processor should play, either in adding more information to the text or in processing it the way it is. The level of qualitative information about the text may vary according to either the visual text planner or visual text processor and is mutually agreed upon.

In other words it matters whether one takes Line A or Line B and the choice needs to be made explicit, from the beginning. In some cases a high level of qualitative information may be useful while in others a lower level of qualitative information may be enough.

If one takes Line B, by adding qualitative information both on a global level – signs - and on a local level – symbols -, one may feel that general information is no longer useful, because such information has been either explained more thoroughly or changed in the course of the process. A canvas may then be absorbed and the text may merely carry signs. There will be no loss in the qualitative information but more specific information will be provided at each given stage.

3.4.3 Visual text programming

The use of visual tools in text planning and processing may be defined as being transparent for two reasons:

1. the visual tools and icons are kept visible, that is, transparent both to the visual text planner and the visual text processor;
2. the visual tools and icons make the text transparent by allowing visual text planners to show what they have planned the text to be and the visual text processor to see what would otherwise have been invisible or subject to arbitrariness.

Textual canvases, textual signs and textual symbols represent qualitatively different instructions for the visual text planner and therefore convey qualitatively different messages to the visual text processor. They are tools for making visible what is not visible in text at a purely linguistic level.

The textual canvases show how a text planner may visually plan and organize texts that have specific communicative purposes. Canvases help text planners reflect on and visualize what they are doing as they prepare for and begin to organize a text. CPP-TRS provides the text planner with visual tools that can integrate qualitative reasoning about text within the text itself. The textual signs provide a text with the additional possibility of explaining the communicative function of a text or a part of a text. They tell text processors on a global level what text is about, before they get to the language of the text. These signs are therefore more immediate and less abstract than the textual symbols.

Textual symbols enhance communication between the text planner and the text processor and raise the semantics of text by making communicative intentions explicit. By using these symbols text planners make more evident to themselves what they are trying to accomplish in the text they are planning.

There are two kinds of textual symbols. The first kind characterizes the style or type of text, while the second kind facilitates an interaction between the text planner and the text processor.

A CPP-TRS-based visual programming of text is founded on a highly skilled interpretation process as icons and schemes represent interpretative processes acting on natural language and on text configuration.

Pictorial representation and pictures in general are most effective in programming, as pointed out by Shu (1988):

1. pictures are more powerful than words as a means of communication; they can convey more meaning in a more concise unit of expression;
2. pictures aid understanding and memory;
3. pictures may provide an incentive for learning to program a text;
4. pictures do not have language barriers and if they are properly designed can be understood by all people, regardless of the language they speak.

Visual text programming, reframed in our terms, uses meaningful graphic representations in the process of programming a text, where programming means specifying a method for explaining those interpretative cues that will allow any consistent text operations to take place.

It is important to remember that even if our minds have powerful recognition capabilities, they are still very susceptible to information overload. It is very important that the graphics are not so overwhelming that they may no longer be processed effectively.

It is evident that the model here presented is intended to raise the semantics of text as well as to facilitate text planning, organizing, structuring and processing by making available some commonly shared visual tools that represent and indicate specific operations. Terms such as writer or reader, sender or receiver, have therefore been redefined as visual text planner and visual text processor sharing the consensual system.

3.4.4 Conclusions

This is a new paradigm. It reverses the traditionally accepted way of looking at text structuring "bottom up" by creating a multiple view of text that is "top down".

Visual tools and icons are not intended to be a substitute for words or sentences; they are rather intended to represent invisible aspects of language so they may be added to text, thereby raising the level of the semantics.

Visual text planning in writing and visual text processing in reading may therefore be conceived as extensions of the traditional way of looking at text production and text understanding, by providing a consensual visual system for facilitating both processes in a transparent and cognitively based way.

3.5 A CPP-TRS-based environment for operating text

3.5.1 Introduction

Operation research is the application of a scientific method to the management of organized systems. The three essential characteristics of operation research are system orientation, use of interdisciplinary teams, and the adaptation of the scientific method to the conditions under which the research is conducted.

A crucial aspect of operation research is model construction. Symbolic models are completely abstract and when the symbols in a model are defined, the model is given content or meaning. A system that has the same structure as another, however different the two may be in content, may be used as a model of the other. Such a model is called an analogue.

A special environment for organizing high quality, high speed information is presented here. The same environment may be considered as a guide to the exploration of the structure of the problem identified in order to reveal further possible courses of action.

3.5.2 On text operations

Whenever we talk or write, or more generally, organize language structures in order to be able to communicate, that is, convey knowledge based on certain intentions, we create texts. CPP-TRS (Tonfoni 1989-94) is an integrated environment based on a well-defined and multilayered visual system with different levels of complexity specifically intended to represent and highlight those different cognitive processes that occur when a text is structured. These may be called text operations. A text needs to be considered both on a global level and on a more local one. Operating a text implies controlling the text itself at different levels. Control on a global level means being able to identify the communicative function of a certain text, whereas control on a local level means being able to distinguish the communicative intention that occurs at the paragraph or even at the sentence level.

According to such perspective both the writer of the text, also called the sender, and the reader of the text, also called the receiver, can be defined as text operators (TO). The system may be productively used to identify and visually represent the text operator's global and local patterns of communication. A global communicative pattern (GCP) means the overall communicative function of a text. A local communicative pattern (LCP) means the intention of the paragraph or sentence as well as the turn-taking that may be requested of another text operator. In some ways communicative patterns are more important than words and sentences because they actually control meaning at a higher level. They are usually apprehended only after processing and interpreting the whole text, which involves time and effort on the text operator's part and many arbitrary decisions. In many cases, those patterns are not found altogether because even text operators may not be able to define them if they have not become aware of them.

CPP-TRS is a well-defined visual organization system that is as consistent as and no less significant than the system of punctuation. In order to be able to operate text consistently, text operators will have to share the operation system both at the level of text production and at the level of text understanding. Text production is defined as text actualization (TA) whereas text understanding is defined as text sensing and perceiving (TSP). To analyze the overall process let us now look at the global dynamic scheme, which represents a special environment for operating text, illustrated in Fig. 3.76.

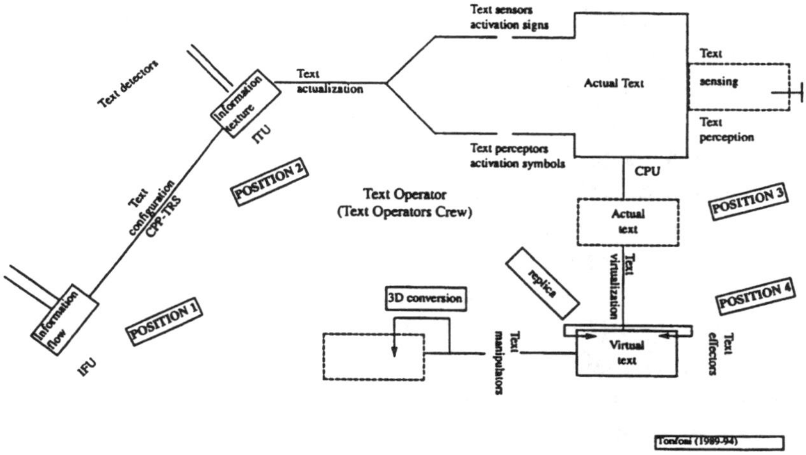

Figure 3.76. Global dynamic scheme of the environment for operating text

First we should observe that the text operator or crew of text operators is placed in the centre of the environment to indicate that this is where the global control is located. In other words, the control of the overall process both at a local and at the global level is with the text-operators' crew (TOC). Now let us proceed stage by stage through the different positions:

Position 1 (information flow): when open, the information flow, made of unprocessed sequences of information, is activated; its basic unit is the information flow unit (IFU) which has not yet been processed. The information flow will undergo a process of texture configuration organized by the text-operators' crew, which means that there will be a visualization of information distribution, made possible by the activation of texture detectors (textual canvases) by the text-operators' crew.

Position 2 (information texture): once configured the information texture can be subdivided into information texture units (ITU). This is the departure point for text actualization. The text actualization process suggests that the text-operators' crew, who have been monitoring the texture configuration process, will be able to convert the information texture, that is those ITU that have been chosen, into an actual text. In order to turn the non-actualized texture into an actual text, the text-operators' crew will have to activate both text sensors (textual signs) and text perceptors (textual symbols).

Position 3 (actual text): once actualized, the text shows the active sensors and perceptors, which act as macro-instructions (sensors) to extract the GCP from the text and as micro-instructions (perceptors) to extract the LCP from the text. As a result of the consensually shared system, text sensors will be able to produce text sensing and text perceptors will be able to produce text-perception. Text sensing means that the text operators' crew will be able to recognize the GCP within the text, indicated by the signs system. Text perception means that the text-operators' crew will be able to recognize LCP within the text, indicated by the symbols system.

CPP-TRS visually processed text units are also defined as communicative patterns units (CPU), which incorporate text and both sensors and perceptors. This means that each CPU is transparent to any text operator based on the same system of representation. The actual text may undergo a further process of reformulation which means that each CPU may be expanded into a virtual communicative space.

Such process is made possible by using and applying the system and is called text virtualization. The text-operators' crew will then operate on text-sensors by modifying their communicative function (trajectory). A sensor that is being modified in its original trajectory is named text effector. The text-operators' crew may also operate on text perceptors by modifying their communicative intention and turn-taking (energy). A perceptor that is being modified in its original energy is named a text manipulator. The result of such a set of operations is a virtual text.

Position 4 (virtual text): Once virtualized, the text shows both the active effectors and the manipulators, which act as macro-instructions (effectors) to modify the GCP of the actual text, and as micro-instructions (manipulators) to modify the LCP of the actual text.

As a result of the consensually shared system, text effectors and text manipulators will be able to modify (virtualize) text sensing and text perception. A virtualization process, where text effectors and text manipulators are active, may be defined as a multilayered conversion.

There is one more possibility in virtualizing text which is to amplify it into a 3-D communicative space. Such process is called 3-D conversion and it is activated by textual objects and textual machines.

To summarize, we can say that text virtualization means the following: The text operator, after having sensed and perceived the actualized text, will decide to operate on the patterned and perceived text to extend it into a communicative environment other than that in which it exists by applying some CPP-TRS components and leaving a trace. Text virtualization means that different communication spaces for an actualized text have been exploited and that a text has become a virtual text with a set of visible operations. It is very important to distinguish between information texture which is not virtual but potential and not yet actualized, and virtual text which is an extension and expansion of an actualized text within a virtual space of communication.

Let us now go back to the overall process represented in Fig. 3.76. This shows that a text operator or a crew of text operators may act at different stages but keep overall control based on a mutual and consensual sharing of the system. The behaviour of a text operator affects the communicative environment. Since the text operator may exhibit many kinds of communicative attitudes, it is very important to be able to recognize and name them consistently in order to have transparency and over all control.

This is why the system is first aimed toward promoting self-awareness in each individual (TO), in a crew (TOC) and within a more extended communicative environment (CPP-TRS-based environment). We all know that the punctuation system allows a reader to distinguish an interrogative sentence from a declarative one. In a similar way, by self-reflecting on communicative behaviour the text operator will be able to identify and use the most appropriate visual representation provided in order to make GCP and LGP explicit. In some ways, CPP-TRS can be defined as a communicative traffic control system enhancing the active role of the text operator. As with any system, by knowing the basic rules, an infinite number of operations may be performed and understood in a quick, unambiguous and easy way. CPP-TRS allows any text operator to actualize any kind of text by following very simple instructions leading to more complex explanations. By conveying GCP and LCP visually, even interactions occurring in different languages at the same time can be facilitated.

The whole system is based on the concept that you may enhance and mutually control communication by visually processing text. The starting point is a visible global organization of text, and the ending point is the actual language of the text. This approach is soundly grounded in cognitive research that is explicitly aimed at explaining the complexities of how our minds apprehend, process and communicate knowledge.

3.5.3 On implementing CPP-TRS-based solutions

Let us now move into a more detailed description of the different components by first identifying the overall purpose of the system, which is presented in Fig. 3.77.

CPP-TRS is for:

Improving textual resolution
Promoting active sensing and perception
Enhancing visualization of information distribution
Increasing speed of knowledge processing
Increasing reliability in positioning detection by communicative pattern recognition
Increasing reliability in information structuring active text sensing and perception

CPP-TRS is also for:

Active processing and utilization of information redundancy
Active processing and utilization of contradictory information
Active extraction of abstractive knowledge
Active high speed and high accuracy sensing and perception
Active knowledge acquisition and noise reduction
Active processing of uncertain information and solving of ill-posed problems
Active determination of sensing and perception strategies.

CPP-TRS (Tonfoni, 1989-94)

Figure 3.77. Purpose display of the overall system

In CPP-TRS there is no such thing as mere language organization, where words and sentences flow out automatically and unambiguously convey intention and meaning, because this does not actually happen. There is rather a process of communicative labelling that makes transparent those aspects of language that are most difficult to represent, which is what actually happens in communication.

Language is very complex. For example, it is very difficult to understand from the sentence "It's cold in here. The window is open" the message "Please close the window". There is no evidence for it whatsoever in the language itself. There is simply no direct link between words, meaning, function, intention and turn-taking. This is the kind of ambiguity we find all the time, because it is implicit in language as a communicative system.

The traditional process of text production is here reversed. Instead of starting with words, producers of a text must first understand their own intention, then design the structure of the text, and finally select words and sentences for it. This approach avoids the pitfalls of the traditional approach by making communicative intentions explicit at the very beginning.

Let us now have a look at the different components shown in Fig. 3.78.

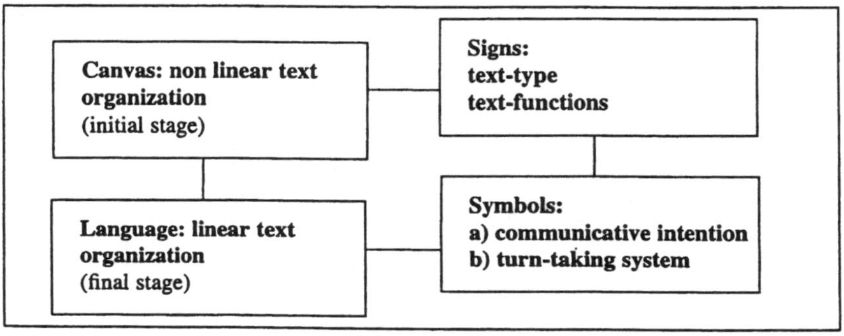

Figure 3.78. Visual text actualization, sensing and perception

The system is being used by the text-operators' crew in a transparent way. The transparent use of the system includes the visual representation of signs and symbols in the text so that the text-operators' crew share tasks by mutually and consensually referring to a common system.

Now let us consider the activated system both in text actualization and in text virtualization by looking at Fig. 3.79.

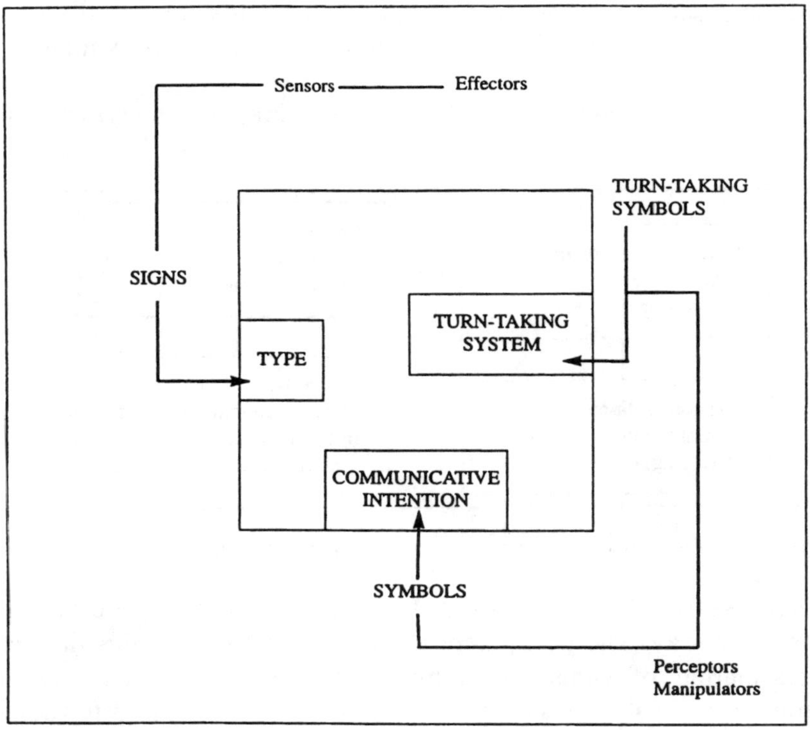

Figure 3.79. Text actualization [sensors] [preceptors] – Text virtualization [effectors] [manipulators]

What has been described up to now is an overall view of a special environment for operating text at different levels and in different stages. The text operator or text-operators' crew is designed to play an active role in the overall process through a defined set of active communication displays.

Through the different components, we may describe the design of a multilayered tool for information detection and transformation, called an active information scanner, which can be activated in positions 3 and 4 of the special environment for operating text. The active scanner design can be seen in Fig. 3.80 and in Fig. 3.81.

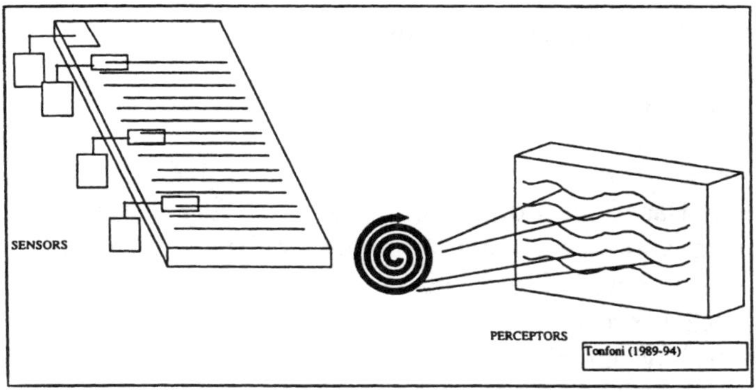

Figure 3.80. Active scanner design for text actualization

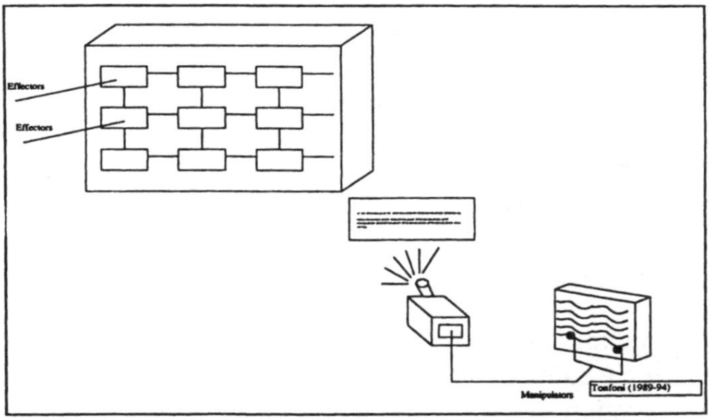

Figure 3.81. Active scanner design for text virtualization

3.5.4 Conclusions

We may conclude that CPP-TRS is both a methodology and a language. Through the methodology those invisible aspects of communication are identified and represented by the meta-language, which complements natural language. As a language by itself, CPP-TRS can be applied to a variety of fields where accuracy of interpretation is of crucial relevance.

3.6 A CPP-TRS-Based E-Mail System Architecture

3.6.1 Introduction

A new model for conceiving and organizing e-mail exchanges is presented here. Based on a definition of qualitatively different categories of messages and networking attitudes, multiple tasks and multiple viewing architecture is illustrated.

CPP-TRS (Tonfoni 1989-94) may be defined as a communicative traffic control system aimed toward facilitating message production and delivery by visually coding and pre-interpreting the message. As in any language, by knowing the syntax an infinite number of sentences can be generated, processed and understood in a quick, unambiguous and natural way, the system here illustrated allows the user to generate any kind of message from very simple information storage to more complex information, filing updating and forwarding.

In the present chapter only those visual schemes and conceptual tools will be presented that are directly relevant for the basic actions and processes that take place in e-mail.

3.6.2 Communicative semantics of a text

The whole system is based on the concept that communication may be enhanced by visually perceiving text. The starting point is a visible global organization of text, and the ending point is the actual language of the text. This approach is soundly grounded in cognitive research that is explicitly aimed at explaining activities going on in the mind (Minsky 1986; Schank 1988). What is intended by "communicative semantics of a text" is the following kind of information about the text, which can be organized at four different levels:

First level: global communicative plan of a text, which involves the kind of operations and actions that work on a cognitive level to begin organizing a specific text.

Second level: global communicative function of a text, which is the overall typology of the text, and the kind of communicative functions the text is supposed to play.

Third level: local communicative intention of the different parts of a text, which means the variety of intentions and styles that can be found in each sentence or paragraph.

Fourth level: local communicative turn-taking of different parts of the text, which means that parts of the text may explicitly require some change or update on the text-receiver's side according to a request made by the text-producer.

Communicative semantics is therefore a qualitative description of quantitative information, whereby qualitative description of communicative function, intention and turn-taking is considered and quantitative information about sentences and paragraphs is as well.

CPP-TRS is therefore an unambiguous, quick, and effective system for reinforcing natural language use in e-mail and net-working, as it is a communication traffic control system designed to facilitate text production, delivery, interpretation, passing, filing and retrieving.

It has been previously mentioned that the system consists of two integrated parts: the methodological component of the system enables receivers to understand how senders have positioned themselves with regard to certain information. It is a complete program that provides visual schemes, models, and tools to organize information effectively.

The visual language component is strictly integrated with the methodology and is the corresponding way of representing those cognitive processes and communicative actions that have been previously identified.

CPP-TRS starts from a global perspective, reflecting visually the communicative intention and function of a message as well as the role the receiver should play by receiving and returning the message. It provides the user with the set of visual tools that are suited to structuring text and communicating effectively. The model and the various applications of it (Tonfoni 1989-94) have been thoroughly analyzed and illustrated in much detail in a very wide literature.

Just as an artist chooses a canvas with an appropriate texture, a sender needs to decide how much "texture of information" the text will have. The texture of text has to do with the density of information conveyed by it and depends on the particular com-municative purpose of the text. Before organizing a message, the sender prepares for it by thinking about the text. A text may be written from a multitude of perspectives and points of view, with many different focuses, intentions and organizational considerations. Each way of writing text enables a different viewing of it and is an invitation to the writer to "redraw" it.

Visual symbols are like road signs that tell a driver what lies ahead or like the notation that directs a musician to play a movement slowly and softly or briskly and lively; they enhance communication between the sender and the receiver of a text. These symbols raise the semantics of text by making its meaning explicit. By using these symbols senders make more apparent to themselves and to their receivers what they are trying to accomplish in the text they write.

In e-mail applications, the first kind of signs and symbols characterize the style or type of message, while the second kind of symbols facilitate an interaction between the sender and receivers of text. The text-style symbols are describe, define, narrate, point out, explain, regress, reformulate, synthesize, analyze, express.

These text symbols are named as verbs, whose precise meaning is based on ancient Latin words that carry strong suggestions about the cognitive process the symbols represent. For instance, describe comes from the Latin word "describo", which means to write about or write around in a free, unconstrained manner. A description can provide as much or as little information as the writer chooses, and it does not need to be put in a logical or chronological order.

The text symbols are designed so that they intuitively convey the intention of the text with which they are associated. For example, the symbol for describe is a spiral with an arrow on the outer end and conveys the free, unconstrained characteristics of a description. Text symbols control text locally. A second set of symbols are the turn-taking symbols. A sender can place turn-taking symbols within the message to facilitate a dialogue with the receivers of the message. These symbols explicitly tell the receivers how the sender wants them to interpret the message and when it is their turn to take part in modifying the message itself.

Signs control the text on a global level. They tell readers on a global level what the text is about, even before the readers begin to read the language of the text. These signs are therefore more immediate and less abstract than visual symbols are.

As in a polittico painting, a sender may shape, partition and organize text to help receivers visually understand its main focus. The sender can put textual elements together so that the arrangement itself emphasizes what is central and what is peripheral — that is, what is most important and what is less important, but still relevant. The different shapes of textual elements may, moreover, provide multiple views of the text as a whole, to enable readers to com-prehend it in all its complexity, and possibly choose and substitute other texts.

In order to plan an overall e-mail system architecture to provide message packaging, filing, storage, retrieval and delivery capabilities, it is important to think about different tools with their different purposes. Users may be merely readers or active commentators; the key role is played by the user, who is perhaps better called the chooser of those different tools that are available.

Users have therefore been redefined as the choosers, to underline the active role they are supposed to play both in deciding the kind of communicative tasks they are addressing and the kind of tools they will choose to operate.

CPP-TRS provides the chooser with a whole set of visual tools that have been conceived and planned to structure text according to an extremely powerful communicative semantics, which allows the text to convey visually those aspects of language that are not conveyed linguistically. The message organizer will be able to process and make transparent those aspects that are most difficult to grasp and therefore to transmit.

The e-mail system thus will provide the chooser with a set of visual tools, shown in Fig. 3.82, that are specifically suited to structuring text and communicating it effectively.

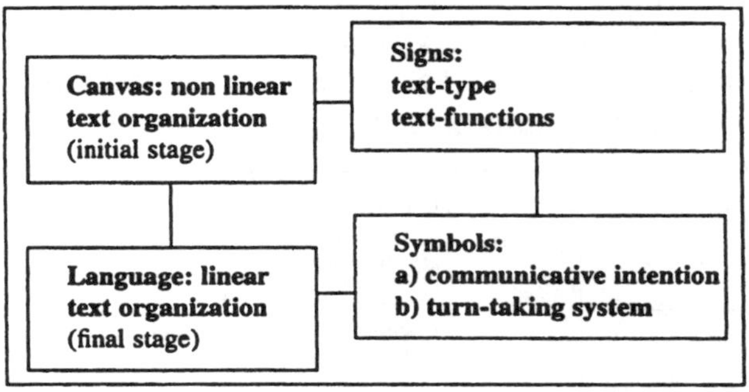

Figure 3.82. Set of visual tools for organizing communicative semantics

What Fig. 3.82 shows are the different kinds of visual tools provided for choosers to organize the communicative semantics of their messages by raising the visual perception of their text.

Starting at the canvas, which is intended to represent the global communicative plan of a text - first level of communicative semantics -, the chooser moves on to signs, which represent the global communicative function of a text - second level of communicative semantics-.

After signs, the chooser finds two kinds of symbols representing local communicative intentions - third level of communicative semantics –, and local communicative turn-taking respectively - fourth level of communicative semantics -.

Finally and at the end result of the previously organized levels, the chooser moves to a linear organization of the text. A large and extremely interesting literature on e-mail has explored internetworking and organization problems at various levels (Ginsberg 1986; McQuillan and Walden 1977, Partridge 1986, 1988; Tannenbaum 1987).

3.6.3 A CPP-TRS E-mail System

The actual approach to e-mail system architecture is based on the identification of a set of qualitatively different communicative actions that refer to message categories and corresponding networking attitudes. Message is defined as "that text or text unit which is meant to move from a certain original location to another targeted one or more and which leaves a trace in the original location in the form of a copy or comes back partially or totally modified in one or more stages. The message is subject to change, revision, multiple agent bunching."

The categories of messages that correspond to qualitatively different attitudes to networking are represented in Fig. 3.83.

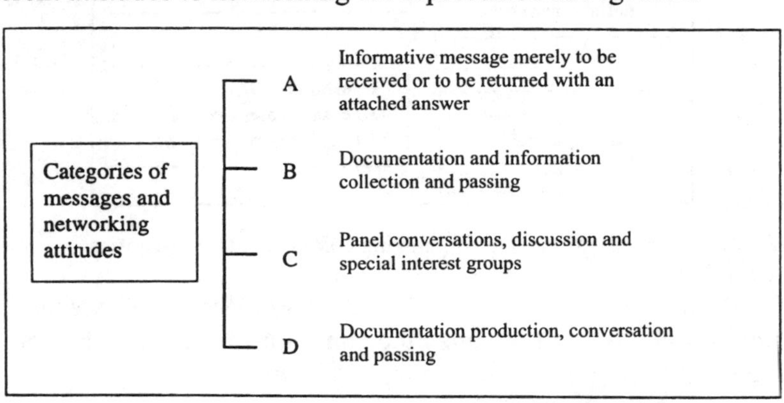

Figure 3.83. Categories of messages and networking attitudes

Let us now specify more precisely the kind of communicative actions that are actually involved in each of these categories, represented respectively by A, B,C and D.

A. By informative message production and delivering, which is merely received or returned with a simple answer, we mean meetings, announcements, single appointments, single news items or updating. Such messages usually move from an individual who takes responsibility and initiative to another individual or more individuals.

B. By documentation and information collecting and passing is meant the conversion of personal files into public documents. The process originates with individuals but is later extended to a collective group. Everyone is able to contribute by expanding text quantitatively, that is by adding more information, and qualitatively by changing qualitative aspects of the text. Information and documentation collecting and passing usually moves from single individuals to the group.

C. By panel conversations, discussions and special interest groups is intended those groups of individuals who are linked together by common tasks or interests and act collectively by playing specific roles, which may change over time but may always be identified. Texts that are produced or passed may be collective news or comments on topics raised and discussed. The group is organized and assisted by a facilitator, who assigns tasks and directs the overall action by assigning turns to participants.

D. By documentation production, conversion and passing what is meant is conversion of existing documents as well as updating and revision. It involves groups of individuals who cooperate by exchanging roles, but are still able to keep track of the kind of communicative operations that are performed on a qualitative level. A directive role needs to be played by someone, even if this is done on a turn-taking basis and is temporary.

The chooser who organizes the Informative Message [Category A] is provided with two specific visual tools: the textual miniature and the canvas for text rotation.

Let us look at the textual miniature. A textual miniature is a visual tool for both individual and collective organization of information to be sent or expanded and completed. In the architecture of an e-mail system here illustrated a textual miniature is suited to the circulation of information on a one-to-one basis. The message sent may also be collective, but it originates with one person and is disseminated to several receivers.

A text miniature is based on a consistent text-partitioning map of the textual space. The text-partitioning map defines text "islands" that contain different kinds of information as shown in Figure 3.84.

The main part of the text space contains what the sender plans as the common and shared part of the text. The other parts of the text space have different communicative functions. The upper part contains the topic and the receiver's name and may also present the perspective suggested to the receiver. The topic islands may also contain specific requests for completion.

The sender may ask the receiver to add some information or merely to acknowledge receipt: some parts of the text may also be highlighted so that the receiver may focus on them. Once highlighted, the receiver may transfer, repeat or headline a word or sentence on a side-island note space. The note space, which adds to or expands the information in the main text, will appear on the right side of the main text, since it is intended to be read after it. A word or sentence may also be highlighted, transferred, repeated and headlined on another side-island, which is called the explanation space.

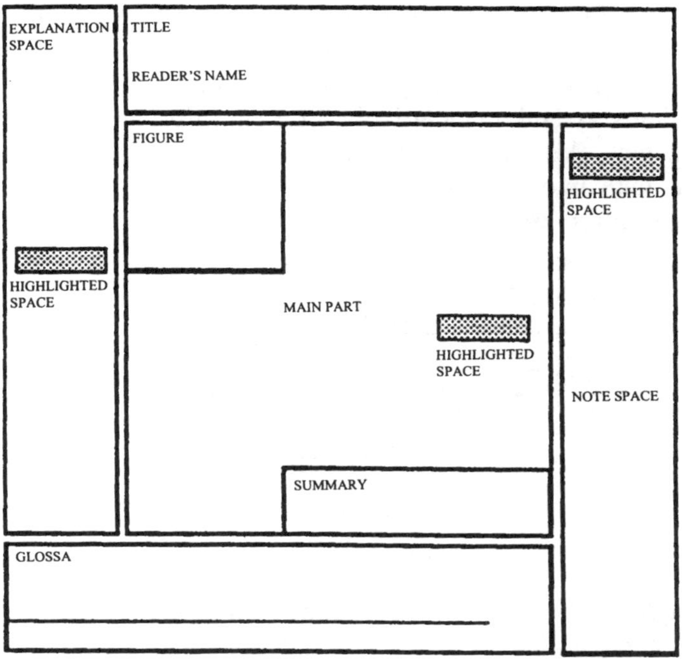

Figure 3.84. Textual miniature structure

The explanation space is intended to provide more information in the form of causal explanation, including the reason something happened, an additional justification or some background information. This space will appear on the left side of the main text, because the sender may like to read it before the main text. The summary space is located at the bottom of the main text; and will contain a summary of the main points or facts that have been presented in the main text.

The glossa space completes the text organization space and appears below the summary space. It may contain comments of the sender but may also be completed by the receiver. The communicative interaction obtained by text-miniaturing is one-to-one and is directed and controlled by the sender; which means that the sender will receive complete messages back from each of the respective receivers.

The receivers fill up different spaces according to the sender's specific request but they do not necessarily know each others' responses.

A canvas for text rotation is suited for the production of many slightly differing messages at the same time. Rotating text is concerned with organizing texts that carry the "same" message to receivers who have qualitatively and quantitatively different information needs. When rotating a text, the sender concentrates on presenting information in ways that take into account a specific reader's viewpoint. Fig. 3.85 shows a visual representation of text rotation.

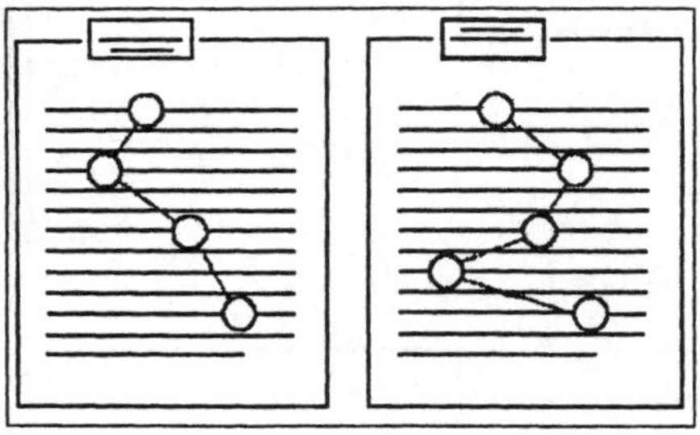

Figure 3.85. Canvas for text rotation

In order to help a chooser collect information, turn it into a document and pass it [Category B], a specific visually organized tool is provided, so called functionalized textual origami.

Rather than considering text in traditional page-oriented fashion, the chooser is challenged to think of text as having a shape. The specific shape given to text visually and explicitly conveys its meaning to receivers. Text can be square, triangular, semicircular or some other shape; there is a syntax of textual shapes because each shape carries a specific communicative intention. Square-shaped text identifies the text as a story, triangular-shaped text tells you that it embodies a memory, and so on.

This syntax of shapes enables the construction of textual structures, as well as the completion of a text. A writer may combine the various textual shapes together to create structures that resemble a textual origami. Receivers may also take their turn in shaping the text in their own ways adding to the structure presented to them.

The text is shaped into various forms according to the specific function of the text. The shape of the text itself therefore provides evidence of the communicative intention so that receivers immediately know the intention.

The system provides a set of two-dimensional geometric signs to shape text. Textual signs are like traffic signs whose specific shapes immediately and unambiguously alert the receiver.

An example of a textual origami to be filled up with functionalized texts or text units is shown in Fig. 3.86.

The sender will be able to organize different parts of the text so that the arrangement itself emphasizes what is central and what is peripheral. The different shapes of the textual elements also provide multiple views of the text as a whole, enabling receivers to comprehend it in all its complexity and, possibly, choose, add, substitute or take away other texts.

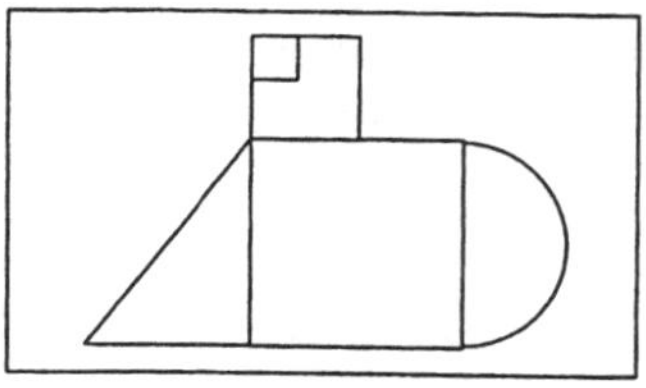

Figure 3.86. Functionalized textual origami

To help groups of individuals who share the same interest to hold panel discussions, by alternating roles [Category C], a specific visual organization is provided, called the textual polittico. A polittico has a central space containing the main subject and a set of complementary parts that can be opened or closed, like doors.

The simplified form of a polittico has three parts - the central main part and two lateral ones - and is called a trittico. Both polittico and trittico are shown in Fig. 3.87.

A textual polittico is particularly suited to text-space organization for raising and discussing a topic. A topic may be raised and receivers and panelists may be required to comment and visually organize their opinions which are then circulated as an expanded text. The open discussion on the topic may be expanded so that more and more receivers are put on the panel. The polittico is different from the miniature because all individuals participating know each others' views and reactions to the topic. Moreover, all parts of the polittico are filled up at a predetermined time so that it is possible to see the views all together.

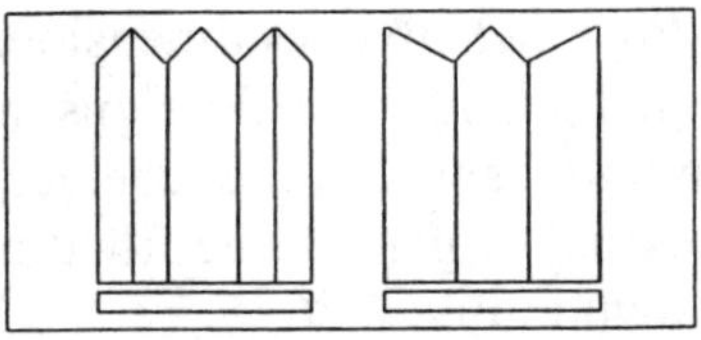

Figure 3.87. Polittico (on the left) and trittico (on the right)

When a chooser initiates a conferencing session, and is the one who decides on a certain number of panelists, assign turns and establishes a deadline so that all the panels of the polittico are completed and viewed at the same time. The initial chooser is the one who raises the topic, but is not the only one receiving comments, as all panelists will know about each other, read each others' comments, communicate with each other and are certainly aware of each others' respective opinions and perspectives.

The polittico is ideal for electronic conferencing among a defined number of panelists. A panel director assigns turns to the panelists and asks each of them to comment on a certain topic once. The panelists know that they have only one turn and are entitled to only one comment on the topic. All of the panelists' comments appear on the lateral sides of the polittico painting - none of them is omitted. The panelists also need to know that someone will value their opinions and draw final conclusions about the panel sessions. The summary is understood as a collectively directed summary. A suitable text may be a controversial topic. It may involve making a decision, solving a problem or asking for opinions.

Panelists must be willing to take a position and make a contribution by stating their individual perspectives. If the panelists would rather debate the issue among themselves, then a polittico structure is not the most suitable form. Panelists need to be aware that their own text is to be a reaction or answer to some triggering issue. They then have to devote time to think about and organize their text to present their own position. The text will be read by other panelists, as well as reported and summarized in the summary space.

A polittico structure involves individually driven collective text organization. Panelists contribute individually - by giving their own personal opinion and by organizing their own texts - to a common project. The panel initiator completes the panel with a collective summary, which is based on reading and interpreting the point of view of each of the panelists.

To assist a group of individuals producing or commenting on existing documents and passing it [Category D], a four-level visual tool is provided, so called textual tapestry. A tapestry is a rich metaphor because it visually carries certain characteristics of collective text organization.

The density of a tapestry corresponds to the quantity of a collectively organized text. Before subdividing tasks, a group must consider how much it needs to be textually organized. Sometimes the quantity of text is externally imposed, such as a specific page limit for a proposal, while other times the quantity is decided by the group. Tapestry also exhibits a complexity of design.

Collective text organization is also complex and must be planned. Some products of a group effort are systematic and structured, while others are more open-ended and unstructured. Collective text organization always requires decision-making about who is going to take care of what and how the collectively organized text is going to come together. Collective text organization also requires cooperation among the group members to produce a shared result.

There are four visual models of tapestries that facilitate group interactions for collective text organization. Each visual model shows a different approach or attitude: the differences in these approaches are indicated visually by the way each model divides and defines space.

Let us now look at Figure 3.88.

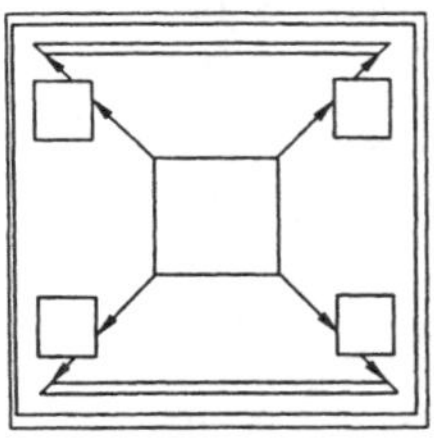

Figure 3.88. Centrifugal model – start from the center and go to the periphery

The model in Fig. 3.88 shows how a group may expand or revise a collectively written text. The text may be something like a report that needs to be updated. This text is represented by the large square in the middle. Arrows point outward from the central square to smaller squares around it. The smaller squares represent the new texts that result from the group's activity. Individuals in the group write these texts as a result of dividing the work and reflecting on or reacting to the original text.

The model is called centrifugal because the group starts from the central text and expands outward to texts written by the group members. This, of course, does not just happen. It must be planned so that the group members know what roles they have in the collective writing process. The group decides how the work will be divided and who will be responsible for writing each text.

The reverse of the centrifugal model is the centripetal model shown in Fig. 3.89.

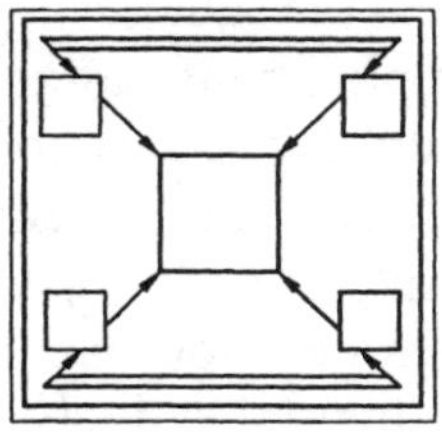

Figure 3.89. Centripetal model – start from the periphery and move to the center

Centripetal means that the process of collective writing flows from the outside parts to the middle. In this model, smaller squares around the outside represent existing texts.

These texts could be a set of individually prepared notes or memos about a certain topic. The group's job is to consider this material and integrate it into a collectively written text, which is represented by the large central square. As in the centrifugal model, the group plans how to integrate the material and makes decisions about the division of work among the group members.

What is interesting to note in both the centrifugal and centripetal models is that they preserve a record of the process of creating the collective text. In the centrifugal model, the existing central text remains with the individual texts triggered by it. Similarly, in the centripetal model the original texts are kept with the resulting integrated text. This preservation of the process of collective text organization is important because it enables us to visualize individual contributions, which will not get lost in the collective process.

The next two models represent a different approach to collective text organization. Fig. 3.90 illustrates the linear rotation model.

In the linear rotation model the group is concerned with organizing a text that deals with a particular topic. Each group member decides to address the topic from the perspective of a particular communicative intention or point of view. For example, one member of the group may decide to organize a descriptive text, while another may want to organize an analytical text and give a more detailed explanation. Other members may choose to organize a topic from the perspective of their particular roles in the group or their viewpoints. What the group is doing at this point is deciding how to partition the space for organizing text. When members finish their contributions, they are passed around and read.

The patterns in the squares are all different because they represent the unique contributions of the members. Each contribution presents a specific point of view about the topic. These may be related to the different roles the members have or to the differences in their individual perspectives or feelings.

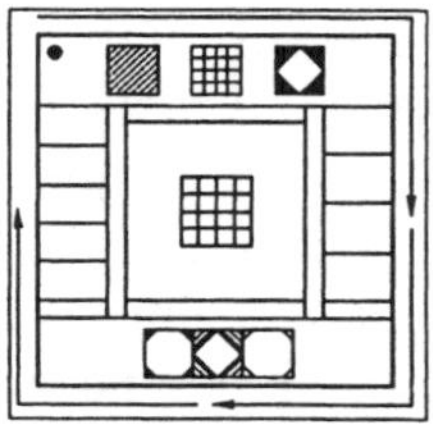

Figure 3.90. Linear rotation model – start from the upper left and proceed around the periphery clockwise

The group then considers each text in a linear fashion, moving around the square space of the visual model, as shown by the arrows. The group then decides which text seems the most relevant; the chosen text is considered the final written output of the group. This text is placed in the middle of the visual model as an indication of its importance.

Fig. 3.91 shows the linear sequential model of collective text organization.

Figure 3.91. Linear sequential model – start from the upper left and proceed across from left to right and then down

In this model the members of the group do not so much represent themselves as they represent a collective thought. The group organizes a collective text. Texts contributed by the group members are represented by rectangles. These texts, as in the linear rotation model, are read and discussed by the group. In this case, however, the order is linear sequential, going from left to right, top to bottom. Because texts are the result of some other group process, they contain systematically defined and accepted material.

As the group discusses these texts, however, controversial issues may surface. Common concerns left undefined or that remain controversial are noted in the spiral motif by a word or short sentence. The group may then view these as questions, problems or issues needing resolution. They may become topics for further discussion.

To summarize, the sender is being asked to start from a global perspective to reflect on the intention and function of each message. A set of "canvases" is provided. These are visual stimuli and global representations of communicative actions. Canvases are visual schemes that describe various communicative processes themselves and are navigation tools to guide the sender through the complexities of transmitting knowledge verbally. Once senders have a global view of what and how they are trying to communicate, they may then proceed to a more detailed structuring of texts. CPP-TRS provides visual signs and symbols for this.

Signs are visual conventions that represent general types of text. Symbols represent communicative intentions and turn-taking. Visual aids such as canvases, signs and symbols that are immediate, unambiguous and consistent, may also be used in combination according to the complexity of the message the sender is organizing.

A global metaphor that has turned into a specific canvas for collective text perception is the collage and mosaic one. These visual schemes are intended to illustrate a collective perception experienced by a group of people writing or reading the same text asynchronously.

The first illustration in Fig. 3.92 helps us visualize a particular type of collective writing defined as collage writing. Collage writing may be done by a group of people who collectively endeavour to create a text by piecing together their individual contributions.

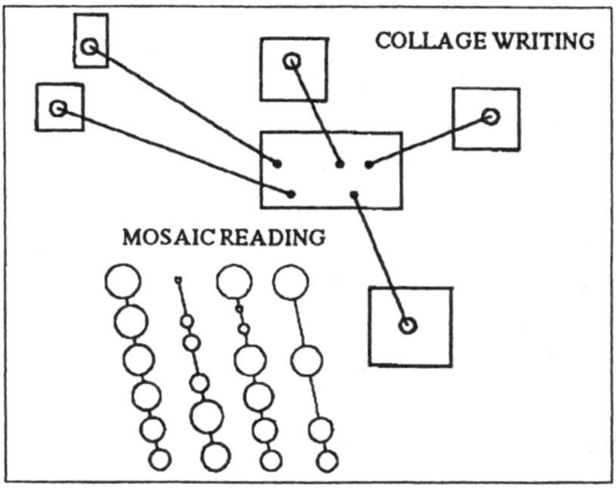

Figure 3.92. Canvas for collective text perception

Initially the group's members get together to discuss the goal of the text, its style and the quantity of information they want it to contain. The members of the group then divide up the work. They discuss who is going to write each part of the text and how all parts fit together. This initial planning process is similar to what a group of architects do when planning the design of a house: the group decides who is responsible for designing the details of each room and makes sure that there is an overall plan so that the rooms fit together into one, unified structure.

Once the members of the group produce their individual texts, they meet to put these pieces together into a single text. This process is visually represented by the top illustration in Fig. 3.92. The five smaller squares and rectangles surrounding the central, larger rectangle represent the single contributions of the individual members of the group. The central rectangle represents the text collectively written by the group.

The group writes a text made out of the texts contributed by the members. As in constructing a collage from various materials, the group cuts pieces of the texts provided by the members and pastes them into the new text they are creating. The way they do this may be somewhat unstructured.

The group thus creates the collectively written text from pieces of the individual texts of the members. This is illustrated in Fig. 3.92 by the lines drawn from peripheral squares and rectangles to the central rectangle. The small circles within the peripheral squares and rectangles indicate that a piece of this text is cut. The point at which a line terminates within the central rectangle represents the place in the collectively written text where the cut texts are pasted. After the whole text is pasted together, the group then collectively reshapes it.

The illustration on the bottom side of Fig. 3.92 represents mosaic reading. Mosaic reading shows how different readers read the same text in different ways. In mosaic reading, a small group of people summarize the same text and then compare how their summaries differ. In the illustration, the line of connected circles at the far left represents the initial text that the group's members read and summarize. The circles depict parts of the text, such as paragraphs. The size of the circles represent the relative importance or quantity of information intended and presented. The three connected lines and circles to the right indicate how three different readers summarize the initial text. As in the representation of the initial text, the size of the circles indicates the relative importance that the reader gives to that part of the text in the summary.

The two illustrations in Fig. 3.92 visually emphasize that a group of people perceive the same text differently. In collage writing, the collective perception of text is different from any one individual's perception of it. In mosaic reading, different readers perceive the same text in different ways. Because of these differences in text perception, it is extremely useful to offer guidance to help the reader perceive text the way it was intended. This is where the signs and symbols come in. They indicate these aspects visually within the text and complement the language.

It is important to agree on the use of definitions such as "mosaic" and "collage", which are familiar terms but, as used here, are precise indications of specific textual operations. This ensures both transparency and evidence. As in a language, syntax needs to be known and respected if an infinite number of sentences are to be generated.

3.6.4 Conclusions

Text composition and execution tool (Tonfoni 1989-94) is the overall structure, meant to represent communicative action taking place interlinguistically and interculturally. It is based on the idea of a shared consensual system that represents invisible aspects of natural language in a whole variety of outputs.

Given the qualitative differences of communication intentions carried through an e-mail system, this system may be a tremendously powerful tool for message filing and retrieval. This is why different networking attitudes have been identified that correspond to different categories of message.

Chapter 4

CTML: a mark-up language for annotating context sensitive documentation

4.1 Introduction

Textual signs and symbols and turn taking symbols have been already introduced and illustrated throughout this book as to indicate how they may in fact represent different styles and stylistic choices in a certain text. The reason why will now illustrate them once again is that they may in fact be considered as a mark up language which may assist information analysts in their interpretation and filing and storing of specifically sensitive information.

So out of the more complex system of visuals, which has been illustrated in each of its components, a highly specific derivative mark up language may be easily conceived. This derivative mark up , also named CTML (Tonfoni 1999) as for Context Transport Mark up Language is in fact explicitly targeted as to make qualitative reasoning in documentation management possible by valuing and enhancing significantly the individual analyst contribution.

Effective performance within an organizational structure is absolutely strictly linked and dependent upon individuals' cognitive ability and teams' cooperative behaviour and communicative competence in monitoring continuously incoming information flows (Tonfoni 1996), some of which are synchronous and consistent, where others are asynchronous and contradictory, therefore in need to be carefully analyzed and filtered.

A consistent visual system for first conceiving and then processing documentation subject to continuous change and also coming in from different cultural contexts here proposed and illustrated. Creating a culture for continuous improvement means being able to value each individual contribution to the production of knowledge within an organization, it is a fact that change may result into a threatening and disrupting factor, if not properly interpreted and handled.

On the other hand, change maybe viewed positively and result in a unique opportunity for individuals to become active protagonists of a most significant process, such as information analysis and information packaging are. Information analysts today more than ever need to be aware all the time that they are working for the present and the future, because what is built steadily is very likely to stay. Documentation, which is conceived as a collective memory based repository of organizational knowledge, is most likely to be productively accessed and reused at different times and for different purposes.

Enhanced data warehousing may therefore be consistently viewed as some kind of high performance knowledge management, where knowledge is the product and result of a set of highly articulated processes of information selection, filtering and then packaging and repackaging, made visible throughout a set of precisely defined documental transitional states, incorporating the history and evolution of each document. Individual analysts' contributions need to be made visible and consistently stored, therefore remaining visible in their own originating context, which may then be available all the time.

Naturally occurring instability and sudden shifts of context and priorities, due to continuous change in the world of information today, may be compensated by a stable interpretive model for documentation packaging, which allows changes and transitional states to be reported, traced back and triggered together with the originating context.

To summarize: each single analyst's visibility will significantly gain out of the use of an enhanced encoding system and out of consistent visualization of the conditions in which the document was first generated and then revised, upgraded, updated or even radically transformed.

4.2 Change: from threat to challenge

Even if interpreted as challenging and rewarding altogether, there is no doubt that change and transitioning conditions, within an organizational structure, may be perceived as threatening and need to be carefully considered and appropriately monitored.

Motivation and commitment by individual analysts or teams to just one stage of a process, which is well known to be at risk and most likely to be either discontinued at any time or passed to different individuals or teams, is hard to keep. Evidently individual analysts involved with a certain process perceive and resent instability either consciously or unconsciously, even if encouraged to feel committed to what they have been asked to do temporarily.

Are results achieved under such conditions actually going to be part of the organizational memory and are they going to be used at all? Which kinds of contributions will actually remain visible if waves of new information, sometime contradictory and fuzzy, are likely to smooth any clear-cut decision making?

These and more questions are likely to come up many times every day within an organizational environment such as the information analysts' community is in fact. Though individuals are encouraged to think of change as an opportunity to learn for themselves first and as a challenge to cooperatively interact and work in teams, insecurity and instability are perceived as diminishing and frightening factors, playing both against personal initiative and collaborative attitudes.

As a consequence of instability, information management subject to continuous change is characterized by the following features:

1. increasingly wide amount of information coming in fuzzy, both synchronously and asynchronously;
2. increasingly wide amount of contradictory and inconsistent data, to be checked and verified continuously;
3. increasingly frequent change of conditions of satisfaction and continuously occurring communicative context shifts.

Decision making based upon instable conditions is perceived as a fuzzy process linked up to concepts such as arbitrariness and chance. Information analysts who are actually the major players as for policy making as they do feed the kind of information makers need as for their daily occurring problem solving, need to make sure they are always able to actually identify relevant information and consequently envision a consistent model for taking action, as to be able to provide consistent advice.

If as for continuous turn taking and rush they are not in a condition to be able to make reasonable projections and predictions, they of course also be inclined, as a consequence, to resist full involvement and full commitment. Decision making, subject to change of roles played by single information analysts and teams all the time, may come out fuzzy and blurry even if unintentionally.

If information analysts who have been assigned a certain task may not foresee how long and to what extent they will actually be left in charge and be effectively responsible for a certain process they have initiated, they are more likely to resist providing their own experience and those knowledge based interpretive clues they have themselves developed in the course of the years, as to facilitate problem solving by someone else, who may come in unexpectedly and under different conditions and within a significantly modified context.

Such resistance is not to be referred exclusively to personality issues and feelings of individual pride, which are obviously there too, it rather comes out of awareness that each change of context, does create a new situation, where a certain problem solving strategy, which may have been previously suggested and considered to be valid, may simply just not work any longer. Documentation management is strictly bound to accuracy, motivation and responsibility on the information analysts' side as they are the ones involved in the process of creation of an interpretive platform to be then handed out to policy makers.

Lack of continuity may radically affect the way information flows and is consequently perceived, monitored and finally packaged. The framework here proposed is aimed toward creating and keeping optimized conditions for positively coping with change and for supporting effective decision making under unstable conditions, by reinforcing the information analysts' role and responsibility, providing both conceptual and practical tools meant to establish most favorable conditions for accurate interpretation.

4.3 Documentation packaging as a teaching and learning opportunity

Documentation (from Latin *documentum*: to be taught) is the output result of whole set of packaging and repackaging operations performed by individuals working collaboratively in groups, within the same organizational structure.

Daily events, current interactions, occasional conversations as well as planned meetings establish the conditions for information to flow throughout different media and in many and various ways and create the context for an accurate understanding in any culture.

Information may therefore be viewed as coming in various flows and waves, to be filtered, categorized and organized as to be accessible and reusable for different purposes at different times. As soon as some information is found to be of relevance as to become stabilized and be turned into a document, then the need to store information in ways, which may be made fully transparent, becomes a major issue. Information packaging based upon topic continuity and context consistency is a most fundamental process, upon which accurate and timely decision making resides.

Availability and accessibility of the originating context in which a certain communicative intercourse actually took place is to be supported by an enhanced labeling system, which may help classify documents consistently and then also speed up effective retrieval. Each document or part of a document, found ready to be linked up to other documents, according to topic continuity and context consistency - will be labeled according to qualitative reasoning criteria upon the nature of information.

Not only is it important to suggest statistical methods for identifying topical words in the form of keywords, but also to add interpretive clues for adding qualitative reasoning on top as well. Really accurate qualitative reasoning upon information is the result of awareness, responsibility and commitment shown by individuals to a certain communicative process or set of processes.

By showing explicitly which kind of information each document contains, will the originating context, in which the information was first packaged become visible. By showing which kind of progressive revisions have in fact be carried on and by indicating explicitly upgrading and updating operations, which have been performed, will it also become possible to trace back individuals information analysts' contributions in revising and repackaging information, all throughout the different context shifts, which have occurred at various times.

Reshuffling pieces of information coming in various forms, as to turn them into highly consistent documents, according to a diversified set of needs and priorities and subject to continuously changing scenarios, entails a specific competence and can neither be left to arbitrary decision making nor left to be the output of some computer program which as for its nature lacks any kind context sensitivity. Not only is a very specific "context sensitivity" required, but also a consensually shared framework for interpretation needs to be available as to be referred to, in order to label documents and pieces of documents consistently, based upon a common understanding and naming of the different operations performed.

Each document, in the process of being labeled and classified as to then be ready to be used for further judgment, represents a learning opportunity for those individuals involved as to discover, define and finally show those very same operations, which have been performed at each given time. As a consequence of explicit illustration of actions taken upon information, the same document may become a teaching opportunity.

According to such perspective, each document will carry its own history attached, while undergoing changes of various kinds, and may therefore be productively viewed as transitioning throughout different states of information, which may, if not appropriately monitored, affect its originating communicative value more or less radically.

Transitional states in a document may therefore be considered those temporarily defined and stabilized states of information, which provide evidence and support for a certain set of decision making processes, which have occurred or are going to occur next. Change applies to the conditions of satisfaction any communicative occurrence entails as well as to priorities and roles played by individuals.

Temporarily stabilized documental states will therefore provide context, evidence and visibility for each individual, who is actively involved with the process. Documentation organized and built up according to such view, does also constitute a tremendously rich repository of collective memories for each organizational structure and does create the context for interpretation and understanding, ready to be in the future as well.

Obviously, a consensually shared encoding system needs to also be made compatible with those information technologies, which are currently used by the organizational structure, as well as made transportable all throughout the different communicative situations occurring in the specific culture, and this is precisely why a common visual system for encoding documentation has been found to be most flexible and handy.

4.4 Enhanced encoding procedures for documentation packaging and repackaging

The process of packaging and repackaging f a certain document may be made visible throughout a consistent interpretive system, meant to describe and define different kinds of communicative actions taken by individuals and groups, as to be finally stored electronically.

Documental operations are here categorized and analytically defined as for the kind of progressive organization, each document is likely to undergo in the course of conversations and meetings, to be finally turned into a consistent attachment to the document stored. Once a consistently interpreted and appropriately packaged document or piece of document is labeled and recognized in its own originating context, it may then be reconfigured many times and more or less radically transformed.

Such process may be activated after relevant clues have been extracted and visually represented by specific icons, supported by the Context Transport Mark-up Language also called CTML (Tonfoni 1996, 1998, 1999), which are of four kinds, and are precisely the following ones:

1. documental signs: meant to indicate the communicative function or type of a document or piece of a document;
2. documental symbols: meant to indicate the communicative style of a document or piece of a document;
3. documental turn taking symbols: meant to indicate the role and interplay between the document producer and the document reader;
4. documental amplifier symbols: meant to coordinate wide sets of documents, which are characterized by topical continuity and context consistency.

By topical continuity, we mean to indicate documents focusing on the same topic, which may be either literally extracted as linear sequences out of the document or abstracted as a result of accurate interpretation and further adaptation at a more conceptual level. By contextual consistency, we mean to indicate documents showing the same communicative context, which may be explicitly declared as to be retrieved easily.

Documental signs, which represent the various communicative functions a document may convey paragraph by paragraph, are the following ones:

Square: for an informative document or piece of a document, which carries information about a specific event or fact, to be linked up with another document or set of documents made available, in order to extend topical continuity and context consistency.

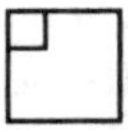

Square within a Square: for a summary of a certain document, which has been produced to reinforce contextual consistency between an original document and its own abstract.

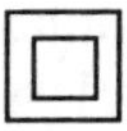

Frame: for a document or piece of document, which is found to be analogous in content to other documents and previously stored cases, is meant to reinforce contextual consistency between and among different documents.

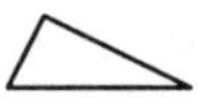

Triangle: for a memory and history generated out of a certain document, meant to establish topical continuity with background information, which has not been previously introduced, because not available in a documental format.

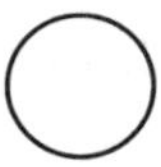

Circle: for a main concept conveyed by a certain document, which has been abstracted as to be linked to other documents showing topical continuity. It is meant to reinforce topical keywords identification and to effectively link together documents, which show the same keyword.

Grouped Semicircles: for main concepts, which are abstracted out of an originating document, and meant to establish both topical continuity and context consistency between the originating document and a set of topical keywords.

Semicircle: for a locally identified concept, abstracted out of a piece of document and meant to reinforce context consistency by establishing further links to other documents, which show the same keyword.

Inscribed Arcs: for indicating the need for an upgrade and update of a certain document; it indicates that a revision process is likely to occur, though it does not declare if such revision will be a major or a minor one.

Opened Text Space: for indicating that an upgrade and update has indeed occurred within a certain document; it indicates that the document has now reached a new revision state as a consequence, though it does not declare if the revision has been a major or a minor one.

Right Triangle: for a comment made to a certain document or piece of a document, coming in a non documental format, where more contextual information is needed, which has to be derived from other external sources, not previously available, based on topical continuity.

Documental symbols are meant to indicate communicative intentions and styles, more locally within a certain document, means sentence by sentence. They are particularly useful as to show contributions made by individuals in the process of creation of a certain document and may be easily incorporated within the final document output resulting, as to provide further interpretive clues, which may significantly add to clarity and visibility.

Documental symbols, which represent different modes of information packaging activated at different times or at the same time, may be combined and used dynamically for repackaging purposes, because they effectively indicate documental transitional states, by declaring explicitly the nature of those changes, which have occurred or are likely to occur next.

Documental symbols are the following ones:

Describe: from Latin *describo*: write around.

It means complementing the original document or piece of a document with as much information as may be found interesting to add without any specific constraints.

It is represented by a spiral, which starts from a central point - the middle point of the spiral indicating the original document - and proceeds toward expanding the document at various degrees, linking it with other documents or pieces of documents or information coming in from different sources and found to be relevant to facilitate the originating document interpretation. In this specific information packaging mode, there is actually no need at all to follow any chronological order; the spiral may be smaller or larger, depending on how much information the document producer - an individual or a team – may find relevant to add.

Define: from Latin *definio*: put limits.
It means complementing the document with limited information about a very defined topic, which has been previously selected and identified as the most relevant one, which is represented by the middle point of the square.
It indicates that there is a specific need to incorporate specific information about a relevant document or piece of a document, which is made available.
Define, means actually describe under specific constraints and implies accurate and most selective focusing on a very limited package of highly specific information.

Narrate: from Latin *narro*: tell the story.
It means complementing the document with various facts and events, which have been referred to in the originating context by following a logical and chronological order.
It indicates a set of major points or facts representing different diachronic stages, which are strictly linked up together in a sequence.
The longer the story is, the more narrative points are actually added according to the document producer's decision making.

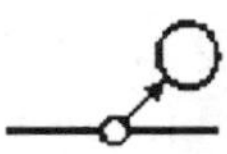

Point out: take a point out of a story chain.
It means isolating a specific event or fact among those reported within a single document, focusing on just that one and adding more detailed information, by expanding it significantly, as to have it linked with other documents, which have been found to be of relevance to that point.
It represents the specific point chosen and does obviously entail the need to look up for more extended information

provided and made available.

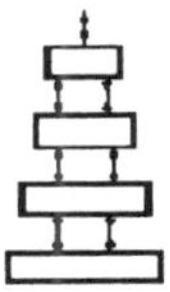

Explain: from Latin *explano*: unwrap, open up.
It means that facts and reasons are given as to support interpretation of a certain event within a certain document or piece of a document.
The document producer may start by indicating the originating cause and proceed toward showing the effects or start with effects and go back to the cause, according to what is found to be more significant.

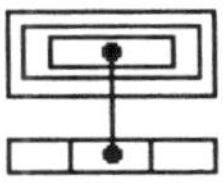

Regress: from Latin *regredior*: go back.
It means that more information about a certain topic, presented within the document, is absolutely needed as to gain a deeper understanding.
It represents a specific topic focusing process and an in depth information expansion, which is activated only for that precise topic.
The document reader may want to consider if further information is needed on that and ask for availability of further resources.

Inform: from Latin *informo*: put into shape, shape up.
It means that any document is the result of some information packaging and that the very specific document indicated is packaged in the most unconstrained way, therefore subject to many and various kinds of repackaging.
It usually leads toward two different kinds of further specification, which are respectively conveyed by the "inform synthetically" and the "inform analytically" indication:

"**Inform Synthetically**" means departing from a larger document or set of documents and proceed toward a summary related to a specific topic, identified as being the most relevant one emerging from the originating document.

"**Inform Analytically**" means departing from a given document or limited set of documents as to expand toward further documents or add more information, which needs to be previously converted into a documental format, which is not available yet.

Reformulate: from Latin *reformo/reformulo*: change shape and shape again.
It means changing the kind of information packaging, which was adopted before and substituting a certain information request with a different one, still related to the same document.
It may turn into a more or less radical transformation of the originating document, according to a precisely defined request or set of requests.

Express: from Latin *exprimo*: push out and press out.
It means adding personal opinions and individual feelings related to facts and events within a certain document; it indicates the most subjective mode of information packaging, which is openly recognized to be bound to a very personal evaluations, judgments and emotional states.

Documental turn taking symbols are meant to define the mode of accessing and reading the document, requested at each given time; they are suggested by the document producer to be followed by the document user; they are the following ones:

Major Scale: it shows that literal interpretation is needed and that those pieces of documents indicated and marked off, should be extracted and quoted literally, the way they were first packaged.

Minor Scale: it shows that accurate interpretation may need a further process of abstraction and that pieces of documents indicated and marked off, may undergo significant reconfiguration processes, up to high level conceptualization.

Open or Unsaturated Rhythm: it shows that accessing the document may lead the user toward incomplete interpretation of those facts and events, which are presented.
It is meant to suggest the user to access more documents and various kinds of sources, which are made available.

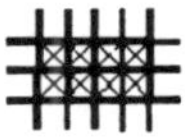

Tight or Saturated Rhythm: it shows that accessing the document will lead the user toward complete interpretation of those facts and events, which are presented. It is meant to suggest the user to stick to the interpretation provided, though access to other sources is still available, as to support evidence.

Documentation amplifier symbols come last and may be added only after the previously illustrated ones have been used; they apply to sets of documents and larger documentation territories and indicate specific operations, which are to be performed as to connect sets of documents, which have been previously encoded and accurately stored. They are the following ones:

Choose: it is meant to represent the dynamic process of first identifying and then deciding between optional contexts for interpretation, which are mutually exclusive, given a certain set of documents.

Identify: it is meant to represent definition of a more specific context, within a broader context for interpretation of a set of documents; it naturally occurs before "search" and "select".

Search: it is meant to represent the dynamic process of choosing among different contexts for interpretation of a set of documents, which are many and compatible as to find the most appropriate one.

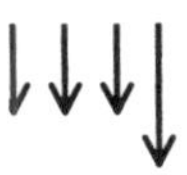

Select: it is meant to represent multiple contexts, which may evolve either synchronously or asynchronously and may be modified once a certain decision making process has been performed, as to be stored and kept as an example.

Copy/Replicate: it is meant to represent the dynamic process of duplication and repetition of a certain context, which, if lost, would radically jeopardize understanding and accurate interpretation of a set of events and facts, described and explained by a set of documents.

→ **Ahead**: it is meant to represent the progression of a certain set of documents, which are linked together by context consistency or harmoniously shifting contexts.

← **Back**: it is meant to represent the need to go back to delete and replace the originating context, which has radically shifted, in the course of various transition states, such that, if not eliminated, would indeed affect consistent interpretation of a whole set of documents.

⇄ **Conflict**: it is meant to represent an emerging inconsistency and incompatibility between various context attributions to a set of documents, which needs to be cleared as to proceed toward any further interpretation.

The document annotation system here illustrated in its various components may be applied at different layers and at various levels of complexity. This system is certainly meant to underline the fundamental role and responsibility of each information analyst who is in charge of encoding the document consistently. The readers are encouraged to refer section 3.1.7 and section 3.1.8 for introductory material on textual signs and textual symbols.

It is obvious that interpretive clues assigned to wide corpora of documents will result into some kind of information mapping system, carrying along its own agenda. Documents, which have been encoded, may then be easily retrieved and most probably will those documents refer to each other so that even a progressive order for accessing and reading parts of them, may be provided. Content continuity as well as context consistency may be made possible in spite of change of time and space.

By retrieving different layers of interpretation of a fact or set of facts, reflecting directly upon stages of information, coming in a documental format, evidence is supported as an attachment for documentation users such as policy makers are. Just like geographic maps only show those features, which become relevant according to the nature and purpose assigned to the map itself, as a political map would of course represent items which would be significantly different than those represented in a geological map, even if of the same region, the same "way of thinking" may be extended to documentation mapping, according to various packaging and repackaging priorities, in continuously shifting contexts.

Energy and time dedicated to quite an expensive process, such context sensitive and enhanced encoding may in fact be, may this way become cost effective. Documentation filed according to such empowered scheme will provide an increased amount of examples and in-house knowledge, which may remain available in time.

Just as any further process of learning within the information analysts community involved may tremendously gain and profit out of such dynamically designed and organized repositories of information, expanding all the time. Feeling of belonging as well as pride are directly connected and linked to motivation and sense of identification and having human experts in charge as for their specialized knowledge accumulated over the time is meant to add value to the whole decision making process.

If it is a fact that not being able to foresee a future reward for a project initiated may be a source of frustration for single information analysts involved, it is equally true that valuing each individual contribution made, by keeping a history of how each information scientists did contribute to it, is a fundamental step as for resourceful rethinking and finding creative solutions to well posed problems.

What could not be brought to completion, as a certain project was disconnected at some point, may turn out to be profitably completed as to be available later on. Just like in the case of a fact, which becomes an example to be triggered back and a learning opportunity also. According to such perspective, individual initiative and authorship as for information analysis and documentation management may be harmoniously incorporated within the text itself as to reflect directly upon history of each document attached, which will explicitly declare roles played by single individuals interpretation providers and information analysts' recognizable teams.

Information technology may be consistently reinterpreted as a powerful carrier, transporting through time and space documentation items viewed as the output result of individuals and teams of experts, whose contributions are made visible according to what they have themselves decided it should be made visible and explicit.

Knowledge repositories built accordingly may in fact encourage active accurate decision making at each given time, ensuring individual who have been in charge of the analysis that each of their respective contribution given will never be lost.

4.5 Conclusions

Reinventing the role of information analysts may result into a very natural process, if memories of single contributors are made available as to constitute an effective repository to be building upon. Inventing throughout continuity with the past turns out to be a highly motivating and reinforcing factor as well as a guarantee against useless dispersion and dissipation of that kind of relevant knowledge, which may only consolidate in the course of many years through the individual and collective effort.

Reinventing the information analysts role culture does not mean reinventing the wheel again and again, pretending it was not there before, it is rather about accurate analysis, rethinking and tracing back of "how the wheel was first invented and then progressively modified and refunctionalized according to continuously changing and diversified kinds of needs as to best suit different kinds of evolving vehicles".

As to extend the metaphor a bit further: innovation within the information analysts community is significantly bound to deeper understanding of the already existing tradition and role and of the reasons why such tradition was first created and evolved in various ways some of which ended up to be diminishing of the individual role even if not planned to be such. And just like new vehicles may only be conceived and designed as a consequence of observation, analysis and most accurate interpretation of the older prototypes, same will be true as for redesigning the information analyst's role of the future, starting today.

Chapter 5

The making of specialized knowledge through metaphors and analogies

5.1 Introduction

The information scientists community is asked to both advance highly specialized knowledge and to make it available to the wider community of information analysts, who may be defined as providers and users of specialized knowledge, in need to access and use pieces of information for specific purposes in their own daily professional life. For example, we all need to know something about progress in research in linguistics, just to be able to ask the right kinds of questions and, even more fundamentally, to be able to describe those communication problems we may encounter and symptoms with "understandable words".

We need to access some knowledge in information science as to be able to define our information needs with accuracy, we need to understand some instructions in order to be able to use the technology at some level of detail and then be able to ask experts more questions, meant to solve more delicate and specific knowledge problems.

Researchers and analysts are therefore facing the need for advancing research and keeping a common code, which may be used by experts in the field in order to communicate, exchange and compare results and to translate some of the results into common sense based explanations, to be then made widely available. In order to circulate new discoveries and highly specialized knowledge and to disseminate findings among a larger community, accurate planning of consistent metaphors and analogies is a fundamental process. Far from being the output of occasional brainstorming, accurate metaphors and analogies come as a result of a skilled art and science.

No metaphor or analogy could represent a specific topic within a highly specialized knowledge domain, without having undergone major processes of redefinition first. This is what will be precisely explored in this chapter: the skilled technique of making metaphors and analogies, which are meant to carry specific meaning and new knowledge, and expected to be understood and absorbed quickly and correctly.

Some specific examples, derived from those metaphors, which have been introduced in the field of Information Design, will be supported and discussed progressively. Metaphors and analogies are part of our life and of our communicative intercourse on a day by day basis. A wide literature has been produced starting from Lakoff's (1980) and continuing with a quite extended literature by Gentner (1983) and Hofstadter (1995), who have been able to harmoniously link knowledge coming from cognitive science and computer science to studies on reasoning in metaphorical processing.

The aim of this chapter is to illustrate a set of conceptual tools which are meant to help the reader analyze the generation, use and interpretation of metaphors and analogies in such scientific research fields as information sciences, and to demonstrate the need for a very accurate framework and to make sure that intended meaning is the one actually being recollected.

In order to be highly reliable, complex processes such as metaphor generation and reasoning by analogy need in fact to be referred to as complex and articulated mechanisms, which need to be described in full details. Metaphors here illustrated have been more extensively explained in a previous book of mine on Information Design (Tonfoni,1998). As for metaphors creation and analogical reasoning, a whole set of processes will first of all be made surface, then visualized and described, in order to grasp the significance of new meaning generation out of old meaning reconfiguration.

To summarize, we will proceed toward an *"in depth exploration"* of the internal mechanisms of scientific metaphors and analogies, those ones, we are asked to think through, in order to be able to recognize and absorb new knowledge in specialized domains and to make it usable on a day by day basis. First of all, we need to recognize that the making of a particularly informative metaphor implies the need to discard many of the originating links; the level of *"informativity",* which means the quantity of information of a certain term, comes in fact as the result of dismissed expectations.

We may therefore add that most powerful and highly informative metaphors come as the result of meaningful links dispersion and new meaningful links establishment; which altogether result into meaning reconfiguration after major semantic shifts have occurred. If you can predict that a certain word may in fact occur in a certain context very easily, then the level of informativity will be low as expectations for such word are high. If you enter a bookstore and ask for a book there is obviously no surprise to it.

If you are entering a shoe shop asking for a book would then the clerk have to figure out what you mean and intend as for this kind of request. And the level of informativity will be higher. What is here presented is a conceptual framework, based upon visualization of those dynamic events, which occur when metaphors are generated out of meaning shifts and progressive meaning reconfiguration.

Before getting into a more accurate definition of what a conceptual toolkit for metaphor generation and evaluation should entail, let us briefly analyze the process of further approximation any metaphor produced to define a highly specialized knowledge domain is supposed to undergo.

In order to undergo major and most significant new meaning attribution processes, will the old term, ready to be metaphorically organized, change its status progressively, proceeding toward different stages, which may resemble transformations occurring when a concept gets transferred from a *"plenum of meaning"* toward a *"vacuum of meaning"*. A *"plenum of meaning"* is precisely that state or space of meaningful links, in which an element represented by a single word is contained at a certain meaning attribution pressure, which is greater than actual contextual and desired meaning attribution pressure.

The whole interpretation space is therefore regarded as filled in with a lot of semantic matter, and full meaning attribution of different kinds including semantic accretion, which may jeopardize new contextual meaning attribution, if a filtering process were not activated to reduce the level of fuzziness, which is still possible.

On the other side a *"vacuum of meaning"* is a desirable situation to be progressively reached, which means the definition of an enclosed interpretation space, from which undesired contextual accretion has previously been removed, so that semantic matter remaining in the interpretation space will exert less pressure than the actually desired contextual attribution, just the opposite from what we would see happen within a plenum of meaning.

A vacuum of meaning represents in fact the state of lowest semantic energy and context attribution possible. Only after a condition of vacuum has been reached, may an accurately selected word be productively turned into an operationally valid metaphor meant to convey very specific meaning. If a vacuum condition has not yet been reached, any word not having undergone a previous process of filtering will carry undesired contextual accretion and fuzziness.

Within an appropriately constructed conceptual realm for metaphor generation and evaluation will each of the words used be first of all transferred and then precisely defined in its own function and range of applications as to be interpreted only after accurate definition, based upon selection and focusing processes, has occurred. This way will the metaphor be properly defined as the *"aggregate of surrounding conditions or influences for facilitating reasoning about a new knowledge domain"*.

As a consequence of a shared core of precisely defined stable concepts, a metaphor designed as for conveying a new concept will be turned into an operational model, where different kinds of operations are made possible and are therefore understandable within a new knowledge domain. In order to represent complex concepts first and to define them afterwards, a combination of different metaphors may work both at a higher speed and at a higher degree of precision and reliability.

A metaphor may in fact cover just some parts and some aspects of a certain domain, whereas a different metaphor may cover some others, which would otherwise be missing or just left out. Though it may sound paradoxical, it is really the discontinuity of meaningful connections that guarantee completeness in conveying the overall meaning of a certain expression.

Metaphorical reasoning is such a complex realm that only a combination of different models for visualizing and expressing various aspects of the word, which is accurately transported from one domain into another one, may allow a map of such complexity to operate in a satisfactory way. We may productively distinguish between *"large scale metaphors"* and *"reduced scale metaphors"*, *"operational metaphors"* and *"narrative metaphors"*, as to be able to accurately evaluate the degree of approximation they may actually reach.

To define a new research field such as the one of Information Design, here is how analogical reasoning and metaphors may help as for understanding the different nature of the new discipline:

"By analyzing carefully new ways to better define a new field such as Information Design (ID) is, there is need to discuss some basic terminology and basic concepts coming from Economics, which may be productively adapted, of course only after accurate reformulation.
To create some common background for discussion, there is a need to create a distinction between Macro Information Design issues just like Micro Economics and Macro Economics. Micro Information Design can be defined as that branch of Information Processing which is in fact concerned specifically with individual information units and knowledge packages. Macro Information Design is concerned with the behavior of aggregate information variables, through the study of information activities and processes of various and different kinds, taken as an interdependent and interconnected whole and dynamically represented in their differentiated stages too".

A process of meaning reconfiguration may reach a high level of accuracy and be related to single words, which need therefore to be redefined according to a preset series of differentiated stages, which are the following ones:

1. *Meaning dissemination:* the process of distributing knowledge widely, meaning in small particles around the old meaning territory.
2. *Meaning dispersion:* the process of spreading information over a wide area after selective separation of relevant from non relevant information has occurred.
3. *Meaning dissipation:* the process of spreading information in random ways to the point of destroying an original meaning territory.
4. *Meaning distortion:* the process of altering or twisting an original and natural condition in meaningful links.
5. *Meaning diffusion:* the process of deliberately spreading information through newly established links according to a preset series of targets.
6. *Meaning disruption:* the process of breaking apart, through an interruption meant to prevent meaning attribution based upon common sense, therefore deeply affecting global understanding.
7. *Meaning distribution:* the process of dividing knowledge, related to various links according to different categories.

8. *Meaning dissolution:* the process of separation between and among knowledge packages, entailing loss of connections, which were previously part of common sense based interpretation.
9. *Meaning disintegration:* the process of separation of the global structure of a term causing transformations at different levels and at various degrees.
10. *Meaning distillation:* the process of extracting abstract concepts out of common sense based knowledge producing therefore the formation of new meaningful conceptual links.
11. *Meaning distinction:* the process of separating different information attributes according to a set of categories after having assigned a new information target.
12. *Meaning dilution:* the process of deliberate reduction of meaningful links out of a given word.
13. *Meaning dissociation:* the process by which a strongly bound combination of words breaks up into simpler constituents capable of recombining under changed contextual conditions.

We see here that what cannot be captured by just one metaphor or analogy is well interpreted by a combination of them, focusing on different features synchronously.

5.2 Models for representing concepts

It is hard to think of new concepts and ways of reorganizing already existing knowledge within a certain scientific domain with no physical models available. Identification of relevant models is therefore important to discover events, enlighten hidden facts and name processes, which would otherwise remain opaque or even unknown.

Since complexity of metaphorical reasoning needs to be handled anyway, being able to refer to a highly articulated framework for interpretation will result in a most useful process meant to facilitate accurate interpretation of newly introduced concepts. Consistently linked up interpretive concepts meant to be used both separately and in combinations, may in fact solve problems, which would be extremely difficult to solve if only the literal level of interpretation was maintained. Introducing new concepts for interpretation of newly observed facts and phenomena will consequently trigger the need for an *"in depth search"* within a new knowledge domain. Applying metaphorical reasoning to newly introduced concepts means *"creating a conceptual platform for supporting new definitions with derivative experience coming from other fields, which are well known and therefore more stable"*.

Such a process may be of tremendous assistance for facilitating reasoning, once the interpretive perspective and point of view have been accurately defined, and may therefore constitute a common background to share new common sense reasoning trajectories. Common sense is most relevant for rethinking and reorganizing of the specific knowledge domain among different individuals involved with cooperative research work worldwide.

Accurate rethinking of the inner nature of each word and consistent redefinition, by introducing physical models of reasoning which can be easily shared, is a fundamental step toward better coordination of any new meaning production process, both by single individuals and groups of individuals acting cooperatively, and coming from a different cultural background.

Accurate observation and interpretation of complex situations, through consistent design of precisely defined metaphors, will lead not just researchers in the field, but also everybody trying to understand specialized knowledge to be able to use it and gain a deeper understanding and mastering of a complex knowledge domain.

Planning and further designing a consistent metaphor may result in an extremely delicate operation, which implies a previous selection process, among different possibilities, followed by consistent reinterpretation of relevant aspects and accurate selection of a technically reformulated lexicon, based upon a global reconfiguration of just those elements, which are the relevant ones to be transferred from one well known domain into a newly established one. Time and times of perception as well as space and spaces of perception should also be foreseen and handled in multiple and different ways.

Consequently, each term needs to be accurately redefined: without such previous reconfiguration process, fuzziness and confusion may arise later. A metaphor, which is produced to support redefinition within a specific knowledge domain, is a dynamic structure, meant to be a reliable cognitive tool for supporting description of newly defined problems and for designing ways to solve them consistently.

Metaphors are dynamic and are therefore specifically suited for grasping inherent complexity of the phenomenon or set of phenomena, which they are meant to represent, both synthetically and analytically. It may sound contradictory and unusual to think of metaphors as analytic procedures, since metaphors are primarily conceived and used as synthetic procedures, planned to establish new meaning connections.

By introducing the concept of progressive reconfiguration of a metaphorical structure stage by stage, the possibility of introducing analytical thinking into the planning and designing of precisely articulated metaphors becomes real. In order to better understand which important implications may be derived, let us more carefully consider the nature of a metaphor as the result of a whole process of meaning attribution and of progressive redefinition stage by stage. A scientific metaphor is the result of accurate transferring of a certain meaning, resulting from a selection of relevant features, from one "old" in the sense of well established and known domain into a "new" and less established because still vaguely designed domain.

A scientific metaphor is meant to enlighten first and then describe, define and explain an event or fact or set of events of facts. Metaphors in poetry do not only allow but welcome a whole set of possible interpretations; they are therefore by definition meant to be centrifugal, because they entail expansion of possible meaning attributions toward different directions and paths, departing from the core definition. In order to make sure that a scientific metaphor is adequate in the sense of conveying specifically that kind of meaning, which is intended to be conveyed by the researcher, it needs to undergo most fundamental changes and become centripetal, which means that different kinds of meaning attribution processes will have to converge toward a very precise core definition.

A scientific metaphor is always generated within a relatively stable realm, it comes from a certain region of meaning where different semantic attributions are possible and may in fact occur, but only some may alternatively prevail or dominate and be transferred into newly established conceptual domains.

Metaphorical perception in poetry is in fact based on most welcome and most appropriate occurrence of various meanings attributions. If this is precisely what metaphors are good for and are about in poetry, the same does not apply to the planning and designing of scientific metaphors, which are meant to facilitate highly specific meaning attribution, as to avoid undesired meaning accretion. To really be reliable any scientific metaphor must be conceived as part of a more global framework. By framework we mean a skeletal structure of meaning, which has been previously designated, set up and accurately checked in each of its single parts, which are all interconnected and joint together.

A scientific metaphor designer needs to proceed toward an accurate and highly specific definition of each of those parts as to turn the *"framework"* into a *"domain"*, which is precisely a meaning territory governed by a single overall ruling principle. Only at that point will the metaphorical domain be ready to be turned into an operative tool for a more articulated interpretation and understanding of a complex phenomenon or set of phenomena, through which a whole set of problems may not only be synthesized, but also analyzed and viewed according to a dynamic and multidimensional perspective as to be first thoroughly explained and then practically approached.

5.3 On metaphorical reconfiguration

A conceptual model for progressive meaning shifting from one domain into another, such as the one presented here, needs to be viewed in conjunction with the previously introduced definition of metaphors. It is aimed toward representing different states of a same metaphor, that has undergone radical changes, here defined kernel changes, because of the activation of meaning transformations, up to meaning explosion of different kinds.

Once a certain word has been accurately reconfigured, showing a set of precisely defined meaningful links, and organized within a highly predictable context according to a consistent mapping of expected associations, and based upon common sense reasoning will it then be possible to proceed toward more powerful and radical changes, as it shows in the following set of metaphorical instructions:

1. *Identify the "information territory".*
2. *Visit the "information territory".*
3. *Choose the suitable "knowledge material".*
4. *Choose the available "knowledge material".*
5. *Select techniques for planning and building.*
6. *Decide techniques for restoration and completion.*
7. *Plan accurately for integration of "present" and "future".*
8. *Plan accurately for integration of "past" and "present".*

9. *Plan on space and time needed for future construction.*
10. *Recognize/analyze space and time needed in the already available construction.*
11. *Plan on functionalities to be "built in".*
12. *Verify the already existing and "built in" functionalities".*

Among those radical changes, which create a further layer of interpretation, meaning perturbation may be defined both as the action of perturbing a set of existing links and as the resulting condition of a word within its own natural context for interpretation.

Perturbation may cause core meaning structures to show substantial modifications; according to this view, inner transformations and reactions will occur in predictable chains of easily monitored reactions which may be easily monitored. Transferring a word from a previous and well established context of interpretation into a new and dynamically evolving context for interpretation implies that the word will change its meaning state, according to a set of principles, which have been applied in a more or less powerful way.

Nevertheless traces of modifications are left and permit tracking on details of each transformation, which has occurred. Progressively transferring a word from a certain domain into another one may also be viewed in terms of revision stages at different levels of approximation, until a final domain is reached through a set of precise operations, which will be performed upon the same word.

Proceeding from expected interpretation toward progressive stages of perturbation will result into a whole set of radical changes progressively discarding meaningful links and consistent reconfiguration will show the specific need for accurate multifaceted representation of those very specific explosion and aggregation processes, causing overall meaning to progressively shift and then radically change at the end as a final output result.

In order to capture such complexity, only a very articulated and complex metaphorical structure may serve the purpose. Observation of radical changes occurring at the kernel level of meaningful links will in fact result into a set of different observational stages to be progressively and dynamically represented; no global perception is ever possible because perturbation and radical changes have occurred in different parts of the originating context and have caused complete restructuring at different levels, just as a set of explosion processes would do.

Observation is therefore bound to different stages: moving around old meaningful links, which had been discarded, is an absolute requirement for any researcher and scientist today in order to focus and to actually understand which changes may be made happen and which may be the consequences of such changes. In the process of metaphor generation, no understanding of the whole set of transformations is ever possible at the same time, only from a specific point of view. Observation is in fact bound to moving around and finding the right perspective for viewing and observing each single explosion or aggregation effect.

Details of each meaning reconfiguration may be grasped only if the observational point of view is shifted by moving around the same word, which is undergoing multiple changes toward many directions at the same time. Observation is also tied to recognizing different kinds of shift in meaning which have occurred at different times, so that an overall global perception about something which has occurred and has produced major changes within the original structure may be gathered.

A more accurate perspective or set of perspectives related to different observational points to be identified may be reached afterwards. It will be part of the originating decision making on accurate design and further representation of the metaphor to include and dynamically visualize those most fundamental changes, which have occurred, according to most appropriate points of observation, so that qualitatively different and extraordinary powerful phenomena, which have occurred, may be grasped in all their complexity.

Without any accurate monitoring of the overall process, it would become very hard to identify those relevant features out of the metaphor, which need to be checked and analyzed in order to get a correct perception of the kinds of perturbation, which have occurred. A dynamic model for metaphor configuration is meant to represent at a macro level those same phenomena, which may be represented at a micro level in terms of "chains of reactions consolidating as new combinations of meaningful links in sequences of words, which have undergone metaphorical processing altogether".

The actual need for representing major changes at a macroscopic level is the direct consequence of the specific intent of turning the overall perception of a highly articulated phenomenon, such as scientific metaphor making, into an experience, which may be described and illustrated in details. Incorporating different points of view still within the same metaphor will guarantee a whole set of differentiated perception modes, which would otherwise be lost, to actually take place.

Everybody accessing a scientific metaphor meant to create a shared and new context of meaningful relationships should always be aware of the continuity of those links and of those new connections established. Well formed metaphors are characterized by stabilized context conditions. A major difference exists between a stabilized word within a relative stable domain and a word, which is characterized by stabilized conditions, coming after major changes and perturbation, which may completely modify the originating set of meaningful links and contexts, within a relatively unstable domain.

Accurate checking on newly established conditions, within a metaphor will be needed in order to find out in details about newly established equilibrium conditions both at a local and at a global level. In other words, not only is the point of observation subject to continuous change and movement, but it is also bound to accurate checking of discontinuous conditions for newly established and highly localized meaningful links reconfiguration processes.

Differences among such meaning reconfiguration processes may be as follows:

1. *Restoring processes*: have the original meaningful structure of a word surface and kept as much as possible within an unstable domain.
2. *Restructuring processes*: have the original meaningful structure of a word surface to be completed and expanded within an unstable domain.
3. *Reassembling processes*: have the original meaningful structure of a word surface as to have as many original parts and pieces as possible fit together, according to a newly designed structure.
4. *Reconfiguring processes*: have a strong common sense based knowledge model available as to have a plausible new meaningful setting consistently organized within an unstable domain.

According to different combinations among and between words, a set of possibilities will be made visible, resulting in new but still predictable meaning aggregations.

Such a model reflects different stages, single steps or degrees just like special phases and positions are parts of an evolving process; it also reflects different states, meaning different *"conditions of meaning pressure"*, which may be productively visualized.

5.4 On reasoning by analogy

Reasoning by analogy, within a specific knowledge domain, has to be conceived in terms of harmonious interplay of different kinds of experiences, through consensual sharing of a common terminology and progressive building of a new framework of reference. This way, it will be much easier to enlighten and share output results based upon a terminology, which may facilitate exchange of information and work in progress within a relatively unstable knowledge domain.

We may think of an analogy as a whole system, which has been originally planned, conceived and designed for a certain set of contexts, which may be defined as *"contexts for meaning management"*. Not only should categories be changed to allow for new categories to come in regarding different aspects of specific knowledge, but an overall reconfiguration of functions and contexts needs to be progressively reshaped to allow the analogy to become *"a same structure but different functions system"* at the same time.

New categories will have to be introduced according to accurate decision making, based upon which kinds of specific knowledge domains should be addressed as suitable and more *"relatively stable"*. In order to be able to proceed toward basic reconfiguration of the analogy management system, the originating context will be progressively taken away, to create space for a new context to be introduced and harmoniously incorporated.

Reconfiguration of the system is based upon consensus on newly introduced categories as well as upon agreement about definitions to be used systematically. Just like for a word undergoing metaphorical transport, old meaning concretions, which have accumulated, will be substituted with new accretions based on consensual decision making upon relevant knowledge to be made available.

First of all, any analogy is first of all recognized in its own original context as to be progressively liberated from undesired and inappropriate contextual concretions. Transitional stages have then to be foreseen as to allow reconfiguration to actually take place by progressively discarding the undesirable meaningful links. Progressive discarding of old meaning needs to be complemented by a parallel process of newly defined task assignment, proceed toward accurate segment reconfiguration.

Finally any analogy management process implies a set of different operations to be performed within the context of each original setting to produce a new and precise definition to be made accessible and to be turned into an operationally valid and productive category to be stabilized, commonly shared and productively used.

The first operation consists of most accurate preparation of a stabilized set of categories and of a stabilized body of knowledge, to be organized as the result of accurate decision making by experts within the specific knowledge domain.

The nature of an analogy management system will therefore be *"polyhedral"*, *"multifaceted"* and *"incremental"*. By *"polyhedral"*, we mean that multiple perspectives will be allowed according to what each reader will be able to see through the metaphor, depending on a predefined task or set of tasks, which have been assigned.

By *"multifaceted"*, we mean that different categories of readers may be foreseen or that the same writer may be willing to change role and play different roles at different times, by redefining precisely the kind of perspective adopted at each given time. By *"incremental"*, we mean precisely that different kinds of knowledge and expertise are made available at different levels of access and use and beside that, each knowledge packet is subject to both update and upgrade processes, according to information coming in from different sources.

Appropriate training will have to be provided at many levels and for different purposes as well as guidelines and precise directions for those specialized writers in charge of feeding in new information. Monitoring on the specialized writer's side will constitute a guarantee for maintenance of a common background of shared knowledge. Explicit use of a shared system of reference and language may result in a most useful device for some communicative intercourses. The very same concept of a cooperating team of experts in various knowledge domains mostly resides upon identification of a commonality of tasks and need for sharing experience within a certain information specialists community, in a fast and highly consensual way.

A high level of cohesion is requested for effective information passing and knowledge packaging through conversational intercourse. There is a precise need for identification, belonging and sharing of a common framework for interpretation and labeling according to a consistent model of reference, which may be most profitably embedded and provided within the system and accurately categorized as *"special common knowledge"* in the field.

As soon as upgrading and updating needs surface and become relevant, those very same information analysts involved, have to understand in depth the already existing context, the way it was progressively built up in time, as to create the conditions for accurate interpretation and to be able to interact consistently.

Diversification of various levels of depth in accessing knowledge as well as in making the same resources available at different degrees, according to their readers, needs and tasks, is one of the most fundamental issues to be addressed. Reasoning by analogy also provides support for diagnosing, searching, solving problems, learning and keeping special knowledge and expertise up to date.

According to the complexity of tasks to be accomplished, an analogy management system will serve as an incredibly effective facility. It may unify worldwide and geographically dispersed communities of information analysts, therefore constituting a real point of reference for further experiences to be shared, evaluated and selected, as to be consistently reported and finally incorporated.

By accessing an analogy management system, intended to be also a learning facility, each information analyst will be motivated into collecting more experiences to be filed according to already set up criteria, allowing for further links to be consistently established, kept and transmitted, especially regarding strong connections between and among cases, recursive similarities and strong analogies, to be recognized as acting both on a global and on a local level, case by case.

As a learning facility in itself, any analogy management system will also provide information analysts with accurate models for effective reasoning about cases, which have been found relevant and have been consequently stored.

Some specific categories coming from a specialized knowledge domain may be chosen to be particularly significant as to be transferred into a different domain.

An analogy management system is also meant to facilitate understanding of new concepts, to support redefinition of old categories and reasoning about new possible links, to be productively established between and among apparently disconnected events, which have not been explored. If reasoning by analogy is primarily based upon regressive thinking and by linking "new" to well known and well established criteria and points of reference, therefore increasing synthetic capabilities, an analogy management system is meant to establish still unknown and unexplored connections between phenomena, which have been only partially categorized and have therefore been labeled as *"unexplained"* or *"exceptional"*.

Reasoning by analogy may result in a very significant support for filing and indexing and labeling cases and triggering them back once they have been appropriately analyzed. Accurate reasoning in a specialized knowledge domain is only possible if complemented by a process of in depth search for those relevant features to be first interpreted in their context, as to be then appropriately analyzed in all their relevant features, which may then be linked back to previously examined and filed cases. This is why interpretation turns out to be a precious and highly articulated process, which may require a set of specific procedures to result in accurate explanation.

Specific attention should be paid to the way analogies are produced in the first place: description, explanation and narration provide qualitatively different knowledge mapping facilities.

Specific categories of questions to be asked need to be identified and analysts in specialized knowledge domains may become involved to create a realistic model for relevant question asking, according to a predefined model for knowledge storing.

Teams of analysts may undergo a global rethinking process meant to help them redefine what they have done, researched and observed, not to mention the accurate process of rethinking of their own criteria and conditions for observation as well as objects of observation. Rethinking of strategies constitutes a tremendous support for a unifying model based upon respect and appreciation of diversity of views and perspectives when a final interpretation of a given fact or phenomenon needs to be reached.

An analogy management system will facilitate analogy recognition, ranking different kinds and degrees of analogy, and will also support consistent evaluation parameters by providing a framework for accurate analysis of highly complex settings and situations.

Inferences will be therefore drawn only if supported by adequate cognitive tools explaining in detail those new concepts, which need to be *"architectured"* and carefully *"engineered"*, according to highly visual models of reasoning helping in the planning and in the progressive visualization of context. Diversity in reasoning may be valued as a precious resource for further research, especially in those areas of expertise and knowledge where a lot of unstable data and experiences still need to be interpreted and to be filed appropriately to be consistently organized and categorized to be reused appropriately.

Information analysts in various knowledge domains, supported with a set of cognitive tools, will be able to see the practical advantages deriving from correct identification of problems, followed by consistent definition of appropriate solution and further monitoring. Information analysts will also find precise motivation for contributing with their own experience to the progressive growth and maintenance of the overall specialized body of knowledge.

Decisions will have to be made about which cases, packaged as stories, should become part of permanent memory, as opposed to just be part of episodic memory. Derivative knowledge and reasoning from previous cases constitutes a tremendous support for accurate decision making and a most effective support for establishing consistent links and connections. Visual schemes provided throughout this book may well create a facilitated communication environment for consistent packaging and dissemination of examples, stories and analogous cases. Analogical reasoning as to discriminate between and among cases may also be obtained and different levels of access based on higher vs lower density of information may be provided to different information users.

5.5 Analogical reasoning upon relevant information

Analogical reasoning needs to be based upon relevant information selection and identification. This is a particularly difficult and delicate process nowadays as for the huge amount of data and information available. One more task which is of crucial relevance for any information analyst today is the capacity of discrimination among a whole variety of information sources available at the same time all of which need careful scrutiny and verification.

The process of verified knowledge distribution within a specific knowledge domain, needs also to be precisely designed as to target different individuals involved with the process of specialized knowledge access and use. Careful examination of new knowledge to be then stored and considered to be solid also supports information analysts as for appropriate solution of specific problems, which have been previously searched and defined as most relevant. Well posing of an information search problem being step number one as an ill-posed information search problem may in fact produce disastrous results and trigger analogical reasoning based on facts and elements, which may not have any substantial link.

It is therefore very important to verify that the labeling system provided to a team of information analysts as to classify documents is consistent enough as to allow for accurate interpretation attachment to each of those documents identified and selected as relevant. What final users of specialized information, who are for example policy makers and decision makers are expecting to have is basically an accurate definition of a set of special knowledge domain categories based upon selected background knowledge found to be of relevance, filed according to a very precise technical terminology and complemented with specific information, resulting from a consistent selection process

Teams of information analysts working together on an information problem experts are in fact in charge of monitoring accurately the definition process of each case as well as the corresponding category assignment. It is most important that they themselves proceed toward selecting or building a very precise set of categories and respective contexts for understanding, here defined as knowledge containers.

Teams of information analysts are therefore in charge of the selection and decision making about *"what may be relevant to whom"* and *"what should be maintained"* and "*how it should be appropriately named"*.

If we were to create a parallel metaphor to better describe the process information analysts teams have to undergo, we may add that they are in fact in charge of "information harvesting and relevant data crop collection".

No doubt that they are the ones playing a most important role in controlling and monitoring information both quantitatively and qualitatively. These teams are in charge to fill in the context for interpretation for each given document produced after having identified such context correctly and named it consistently. Only at this very precise stage of the process, may specialized knowledge users really come into play.

According to different kinds of needs and demands they have, specialized knowledge users, who are decision makers have to proceed toward definition of their own information needs both quantitatively and qualitatively and case by case; they will therefore have to get access to specific knowledge after having previously expressed their own requests

After the process of specialized knowledge acquisition has been completed, will it become much easier for policy makers to first make a decision and then to share and pass knowledge around. Dissemination and distribution of verified knowledge at this level will of course entail adequate understanding of a commonly shared terminology and also requires a very active role in the organization, sharing and dissemination of specialized knowledge.

A mark up language such as CTML, which also operate as a text annotation system may analogically be defined as an *"envelope language"*. It will in this case in fact be a transitional language, operating between and among different states of information, as to facilitate specialized knowledge reformulation and final packaging. Highly specific visual tools as those provided in the previous chapters are needed to provide information analysts with the right equipment, which is in fact absolutely relevant to insure success to any knowledge conversion operation, based on monitoring and control capabilities upon the overall process. We can summarize this concept as follows:

> *"Textual scenography is the art of representing a document on a perspective plan reflecting one or more specific point or points of view and not on a principal interpretation axis. It is therefore considered to be the art of perspective representation applied to the organization and representation of stage cooperative working scenarios.*
> *Orthographic projection being the specific technique and set of techniques meant to focus on a specific point desired or set of specific points desired".*

Information analysts cooperating as for packaging consistent bodies of specialized knowledge will also have to be able to accurately monitor the new incoming information as for upgrading and updating already consolidated knowledge based upon new important facts.

Knowledge packages designed as for highly specific needs and demands should by all means undergo a process of progressive reconfiguration according to a real need for additional knowledge any decision maker has and will continue to have even when some decision has in fact been made. New knowledge coming as a result of new cases observation, found to serve as a productive example, may initially come as a whole set of raw new data, scattered information and stories to be filed consistently.

Some of this new information may well serve as reinforcement for precedent cases diagnosis, and may therefore constitute an important platform for further expansion in a whole range of applications.

Out of an already tested and carefully verified procedure, new information acquisition and verification may also provide evidence for newly discovered aspects to be further explored or even open up possibilities for new inferences to be drawn.

New knowledge coming in and new cases supporting evidence will create a significantly wide repository for further search and verification as to be finally selected and consolidated.

5.6 Conclusions

As to conclude let us briefly analyze some commonly agreed upon definitions. The definition of metaphor, introduced by the Webster Dictionary, is as follows: "a figure of speech in which a word or phrase denoting one kind of object or action is used in place of another to suggest a likeness or analogy between them". This short statement is indeed quite interesting because it points out first of all that a metaphor, (from Greek: metaphora, transfer), is a "figure of speech" which in cognitive terms means a visual representation of language activities meant to convey some meaning, which would not be visible otherwise.

A metaphor in today's world of information overload needs not to be based upon a fuzzy model, complexity involved into any communicative intercourse does in fact entail a high degree of accuracy and precision. In order to be properly handled, different layers of meaning need in fact to be first isolated and recognized and defined, to be then prioritized, selected and progressively ordered as to allow for final and most appropriate output result of the metaphor to actually take place.

The concept of *"metaphor performance"* derives directly from the more general concern as for practical consequences any metaphor may in fact have today in the world of specialized knowledge management and information analysis. The visual dimension some metaphors may carry is meant to underline the fact that a metaphor may be easily turned into a conceptual tool for explanation, which may be quite effective.

The Random House Dictionary has a slightly different but completely compatible definition as for metaphor, which is the following one: "a figure of speech, in which a term or phrase is applied to something to which it is not literally applicable in order to suggest a resemblance". The *"literal applicability"* would entail illustration of observable elements, whereas *"resembling"* implies projection of meaning according to more abstract criteria, which means being able to observe through a theory and not just with naked eyes as previously indicated in this book any information analysts should be very skilled at doing.

Not only do we have highly defined metaphors anyway, but also *"mixed metaphors"* called *"similes"*, which are used or regarded as being used to represent "something else still close some" and to be still defined. Analogy is defined in the Random House Dictionary as "a similarity between the features of two things, on which a comparison may be based" and again as "a form of reasoning in which one thing is inferred to be similar to another thing in a certain respect on the basis of the known similarity between the things in other respects".

If we take literally this definition of "form of reasoning" we can see the implications of it, which are somehow related to a visual representation of a complex and dynamically evolving process, occurring in many and various steps. We are the analogies we build, becomes a common sense assertion almost as much as "cogito ergo sum".

We share the memories we have built, and we create "modes of reasoning" brick by brick, where each brick stays for one piece of memory, one single story, each at a time. Out of the many bricks made by stories, will the resulting overall building create the way we see and interpret events and facts in our lives. We will only build upon those buildings and expand on their originating structure: organization of new knowledge and experience is in fact strongly linked and bound to the already existing conceptual structure. As it is so important for building to stand and be constructed properly, the same responsibility is to be assigned to information scientists and analysts who are through their own documentation selection, annotation and knowledge packaging actually creating the building for the present and the future to stand in time.

This book is after all meant to indicate how important and precious their role is and how they need to be valued as for their results. As to get the appreciation their activities deserve they of course have to be provided with most sensitive tools and skills meant to ensure the highest quality of their daily work which is so delicate and meaningful altogether.

Bibliography

Abraham, A., Jain, L.C., and Kacprzyk, J. (Ed.)
2003 *Recent Advances in Intelligent Paradigms and Applications*. Springer-Verlag, Germany.

Ackley, D., Hinton, G. and Sejnowski, T.
1985 A Learning Algorithm for Boltzmann Machines, Cognitive Science, 9: 147-169.

Allen, N.
1994 The Application of Artificial Neural Networks to the Fast Assessment of Transient stability on a Multi-Machine Power System, PhD Thesis, University of South Australia.

Austin, J.L.
1962 How to Do Things with Words. Oxford, Oxford University Press.

Baba, N., Jain, L.C. and Howlett, R.P. (Ed.)
2001 Proceedings of the Fifth International Conference on Knowledge-Based Intelligent Engineering Systems (KES'2001). IOS Press, The Netherlands.

Ballard, D.H., and Brown, C.M.
1982 Computer Vision, Englewood Cliffs, NJ, Prentice-Hall Inc.

Ballmer,T.
1975 Sprachrekonstruktionssysteme und einige ihrer Anwendungsmöglichkeiten in Satzund Textlinguistik, Kronenberg, Taunus Scriptor.

Barr, A. and Fiegenbaum, E. A.
1981 The Handbook of Artificial Intelligence, William Kaufmann, Inc., Los Altos, CA.

Bohr, N.
1922 The Theory of Spectra and Atomic Construction, Cambridge, Cambridge University Press.
1934 Atomic Theory and the Description of Nature, Cambridge, Cambridge University Press.
1958 Atomic Physics and Human Knowledge, New York, NY, Wiley.

Carpenter, G.A. and Grossberg, S.
1987 ART2: Self-organization of Stable Category Recognition Codes for Analog Input Patterns, Applied Optics, 26(23): 4919-4930.

Chen, Z.
2000 Computational Intelligence for Decision Support. CRC Press USA.

Chomsky, N.
1957 Syntactic Structures, The Hague, Mouton.
1965 Aspects of the Theory of Syntax. Cambridge, MA, MIT Press.
1975 The Logical Structure of Linguistic Theory. New York, NY, Plenum Press.
1976 Reflections on Language. London, Maurice Temple Smith.
1977 Essays on Form and Interpretation. Amsterdam, North Holland Publishing Company.

1980 Rules and Representations. Oxford, Blackwell.

Cruttenden, A.
1986 Intonation. Cambridge, Cambridge University Press.

Damiani, E., Howlett, R.P., Jain, L.C. and Ichalkaranje, N. (Ed.)
2002a Proceedings of the Fifth International Conference on Knowledge-Based Intelligent Engineering Systems. Vol. 1. IOS Press, The Netherlands.
2002b Proceedings of the Fifth International Conference on Knowledge-Based Intelligent Engineering Systems. Vol. 2. IOS Press, The Netherlands.

De Beaugrande, R.A. and Dressler, W.
1981 Einführung in die Textlinguistik. Tübingen, Max Niemeyer Verlag.

Dittenbach, M., Merkl, D. and Rauber, A.
2000 Using Growing Hierarchical Self-Organizing Maps for Document Classification. Proceedings of the 8. European Symposium on Artificial Neural Networks (ESANN'2000) April 26-28, 2000, Bruges, Belgium.

Eco, U.
1975 Trattato di Semiotica. Milano, Bompiani.

Fillmore, C.
1977 "The Case for Case Reported". In Cole, P. and Sadock, J. (eds.), Syntax and Semantics. Volume 8: Grammatical Relations, New York, NY, Academic Press.
1980 "The Case for Case". In Bach and Harms (eds.), Universals in Linguistic Theory, 1-88, New York, NY, Holt, Reinhart and Winston.

Geldof, S.
1994 "Report on the 1993 Bolzano International School in Philosophy and Artificial Intelligence: NLP and Multilinguism", in AICOM, 7 (3/4): 227-233.

Gentner, D.
1983 "Structure Mapping: a Theoretical Framework for Analogy", in Cognitive Science, 7 (2): 155-170.

Ginsberg, K.
1986 "Getting from Here to There", in Unix Review, January 1986, 45.

Grossberg, S.
1982 Studies of Mind and Brain, Boston Studies in the Philosophy of Science. 70, D. Reidel Publishing Company, Boston.

Halliday, M.A.K., and Hasan R.
1976 Cohesion in English. Longman.

Hebb, D.
1949 Organisation of Behaviour. New York, John Wiley & Sons.

Hobbs, J.R.
1979 "Coherence and Coreference", in Cognitive Science, 3: 67-90.

Hofstadter, D.
1995 Fluid Concepts and Creative Analogies. New York, NY, Basic Books.

Holland, J.H.
1975 Adaptation in Natural and Artificial Systems. MIT Press, Cambridge MA.
Hovy, E., and Nirenburg S.
1992 "Approximating and Interlingua in a Principled Way", in Proceedings of the DARPA Speech and Natural Language Workshop, Nawsthorne.
Howlett, R.P. and Jain, L.C. (Ed.)
2000 Proceedings of the Fourth International Conference on Knowledge-Based Intelligent Engineering Systems. IEEE Press, USA.
Howlett, R.J. and Jain, L.C. (Ed.)
2001a Radial Basis Function Networks. Vol. 1. Springer-Verlag, Germany.
2001b Radial Basis Function Networks. Vol. 2. Springer-Verlag, Germany.
Howlett, R.J., Ichalkaranje, N., Jain, L.C. and Tonfoni, G. (Ed.)
2002 Internet-Based Intelligent Information Processing. World Scientific Publishing Company Singapore.
Hyotyniemi, H.
Text Document Classification with Self-Organzing Maps, http://www.uwasa.fi/stes/step96/step96/hyotyniemi3/. (accessed 21/3/2001)
Illuminada, B. Barriga, A. Jimenez, C., Lopez, D. and Sanchez, S.
2000 Microelectronic Design of Fuzzy logic Based. CRC Press USA.
Jain, A., Jain, A., Jain, S. and Jain, L.C. (Ed.)
2000 Artificial Intelligence Techniques in Breast Cancer Diagnosis and Prognosis. World Scientific Publishing Company, Singapore.
Jain, L.C. (Ed.)
1995a Electronic Technology Directions Towards 2000, ETD2000. Vol. 1. IEEE Computer Society Press, USA.
1995b Electronic Technology Directions Towards 2000, ETD2000. Vol. 2. IEEE Computer Society Press, USA.
1997a Soft Computing Techniques in Knowledge-Based Intelligent Engineering Systems. Springer-Verlag, Germany.
1997b Proceedings of the First International Conference on Knowledge-Based Intelligent Engineering Systems. Vol. 1. IEEE Press, USA.
1997c Proceedings of the First International Conference on Knowledge-Based Intelligent Engineering Systems. Vol. 2. IEEE Press, USA, 1997.
1999 Proceedings of the Third International Conference on Knowledge-Based Intelligent Engineering Systems. IEEE Press, USA.
2000a Innovative Teaching and Learning: Knowledge-Based Paradigms. Springer-Verlag, Germany.
2000b Evolution of Engineering and Information Systems. CRC Press USA.
Jain, L.C., Chen, Z. and Ichalkaranje, N. (Ed.)
2002 Intelligent Agents and Their Applications. Springer-Verlag, Germany.
Jain, L.C. and De Silva, C.W., University of British Columbia (Ed.)
1999 Intelligent Adaptive Control. CRC Press USA.

Jain, L.C. and De Wilde, P. (Ed.)
2001 Practical Applications of Computational Intelligence Techniques. Kluwer Academic Publishers, USA.
Jain, L.C. and Fanelli, A.M. (Ed.)
2000 Recent Advances in Artificial Neural Networks: Design and Applications. CRC Press USA.
Jain, L.C. and Fukuda, T. (Ed.)
1998 Soft Computing for Intelligent Robotic Systems. Springer-Verlag, Germany.
Jain, L.C., Halici, U., Hayashi, I., Lee, S.B. and Tsutsui, S (Ed.)
2000 Intelligent Biometric Techniques in Fingerprint and Face Recognition. CRC Press USA.
Jain, L.C., Howlett, R.J., Ichalkaranje, N., and Tonfoni, G. (Ed.)
2002 Virtual Environments for Teaching and Learning. World Scientific Publishing Company Singapore.
Jain, L.C., Ichalkaranje, N. and Tonfoni, G. (Ed.)
2002 Advances in Intelligent Systems for Defence. World Scientific Publishing Company Singapore.
Jain, L.C. and Jain, R.K. (Ed.)
1997 Hybrid Intelligent Engineering Systems. World Scientific Publishing Company, Singapore.
1998a Proceedings of the Second International Conference on Knowledge-Based Intelligent Engineering Systems. Vol. 1. IEEE Press, USA.
1998b Proceedings of the Second International Conference on Knowledge-Based Intelligent Engineering Systems. Vol. 2. IEEE Press, USA.
1998c Proceedings of the Second International Conference on Knowledge-Based Intelligent Engineering Systems. Vol. 3. IEEE Press, USA.
Jain, L.C., Johnson, R.P., Takefuji, Y., and Zadeh, L.A. (Ed.)
1998 Knowledge-Based Intelligent Techniques in Industry. CRC Press USA.
Jain, L.C. and Kacprzyk, J. (Ed.)
2002 New Learning Paradigms in Soft Computing. Springer-Verlag, Germany.
Jain, L.C. and Lazzerini, B. (Ed.)
1999 Knowledge-Based Intelligent Techniques in Character Recognition. CRC Press USA.
Jain, L.C., Lazzerini, B., and Halici, U. (Ed.)
1997 Innovations in ART Neural Networks. Springer-Verlag, Germany.
Jain, L.C. and Martin, N.M. (Ed.)
1999 Defence Science and Technology Organisation, Fusion of Neural Networks, Fuzzy Logic and Evolutionary Computing and their Applications. CRC Press USA.
Jain, L.C. and Vemuri, R. (Ed.)
1998 Industrial Applications of Neural Networks. CRC Press USA.
Joshi, A.K.
1991 "Natural Language Processing", in Science, 253: 1242-1249.

Karr, C.L. (Ed.)
1999 Practical Applications of Computational Intelligence for Adaptive Control. CRC Press USA.

Karr, C.L. and Freeman, L.M. (Ed.)
1999 Practical Applications of Genetic Algorithms. CRC Press, USA.

Kennedy, M., Rovatti, R., and Setti, G. (Ed.)
2000 Chaotic Electronics in Telecommunications. CRC Press USA.

Kohonen, T.
1998 Self-Organisation and Associative Memory. Springer-Verlag, Berlin.

Lakoff, G., and Johnson, M.
1980 Metaphors we Live by. Chicago, IL, The University of Chicago Press.

Lazzerini, B., D. Dumitrescu, and Jain, L.C.
2000 Fuzzy Sets and their Applications to Clustering and Training. CRC Press USA.

Lazzerini, B., D. Dumitrescu, Jain, L.C., and Dumitrescu, A.
2000 Evolutionary Computing and Applications. CRC Press USA.

Lehnert, W.G.
1981 "Plot Units and Narrative Summarization", in Cognitive Science, 2: 293-331.

Mann, W.C., and Thompson, S.A.
1987 "Rethorical Structure Theory: A Theory of Text Organization", in Polany L. (ed.), The Structure of Discourse, Ablex Publishing Corporation.

Marr, D.
1982 Vision, New York, NY, W.H. Freeman and Company.

McCulloch, W. and Pitts, W.
1943 A Logical Calculus of the Ideas Immanent in Nervous Activity, Bulletin of Mathematical Biophysics, 7, pp. 115-133.

McKevitt, P. (Ed.)
1994a "Integration of Natural Language and Vision Processing: More Computational Models and Systems", in Artificial Intelligence Review, vol. 8, n. 5-6, Dordrecht, Kluwer Academic Publishers.
1994b Proceedings of the Workshop on Integration of Natural Language and Vision Processing. Twelfth American National Conference on Artificial Intelligence (AAAI-94), August 1994, Seattle, WA.
1994c "Special Issue on Integration of Natural Language and Vision Processing", in Artificial Intelligence Review, 8 (2-3), Dordrecht, Kluwer Academic Publishers.
1994d "Visions for Language", in Proceedings of the Workshop on Integration of Natural Language and Vision Processing. Twelfth American National Conference on Artificial Intelligence (AAAI-94), Seattle, WA, August 1994, 47-57.

McQuillan, J.M., and Waldens, D.C.
1977 "The ARPA Network Design Decisions", Computer Networks, 243-289.

Mealing, S., and Yazdani, M.
1990 "A Computer-Based Iconic Language", in Intelligent Tutoring Media, 1 (3), 133-136.
Medsker, L. and Jain, L.C (Ed.)
2000 Recurrent Neural Networks: Design and Applications. CRC Press USA.
Minsky, M.
1986 The Society of Mind. New York, NY, Simon and Schuster.
1975 "A Framework for Representing Knowledge", in Winston P.H. (ed.), The Psychology of Computer Vision, New York, NY, McGraw-Hill, 211-277.
Narasimhan, V.L., and Jain, L.C. (Ed.)
1996 The Proceedings of the Australian and New Zealand Conference on Intelligent Information Systems. IEEE Press, USA.
Nishida, T. (Ed.)
2000 Dynamic Knowledge Interaction. CRC Press USA.
O'Nuallin, S., and Smith, A.G.
1994 "An Investigation into the Common Semantics of Language and Vision", in The Artificial Intelligence Review, 8: 113-122, Dordrecht, Kluwer Academic Publishers.
Palade, V., Howlett, R.P., and Jain, L.C. (Ed.)
2003a Proceedings of the Fifth International Conference on Knowledge-Based Intelligent Engineering Systems. Vol. 1. Springer-Verlag, Germany.
2003b Proceedings of the Fifth International Conference on Knowledge-Based Intelligent Engineering Systems. Vol. 2. Springer-Verlag, Germany.
Partridge, C.
1986 "Mail Routing Using Domain Names: An Informal Tour", in Proceedings of the 1986 Summer USENIX Conference, Berkeley, CA, 366-76, Usenix Association.
1988 Innovations in Internetworking. Norwood, MA, Arch House.
Previati, G.
1990 La Tecnica della Pittura. Milan, Sugarco.
Quinlan, J.R.
1983 Learning Efficient Classification Procedures and Their Applications to Chess End Games, in Machine Learning, An Artificial Intelligence Approach, R.S. Michalski, J.F. Carbonell, and T.M. Mitchell (Eds.), Toga Publishing Co., Palo Alto, CA, 463-482.
Radford, A.
1988 Transformational Grammar. Cambridge, Cambridge University Press.
RuleBook
1994 EXSYS. Inc., Albuquerque, NM.
Rumelhart, D.E., Hinton, G.E. and Williams, R.J.
1986 Learning internal representations by error propagation, Parallel Distributed Processing: Explorations in the Microstructure of Cognition, Vol. 1., Chapter 8, Cambridge MIT.

Russo, M. and Jain, L.C. (Ed.)
2001 Fuzzy Learning and applications. CRC Press USA.

Sato, M., Sato, Y. and Jain, L.C.
1997 Fuzzy Clustering Models and Applications. Springer-Verlag, Germany

Schank, R.C.
1972 Conceptual Dependency: "A Theory of Natural Language Understanding", Cognitive Psychology, 3 (4): 552-631.
1981 Inside Computer Understanding, Hillsdale. NJ, Lawrence Erlbaum Association.
1988 The Creative Attitude. New York, NY, Macmillan.

Schank, R.C. and Abelson, R.P.
1977 Scripts, Plans, Goals and Understanding: an Inquiry into Human Knowledge Structures. Hillsdale, NJ, Lawrence Erlbaum Associates.

Searle, J.
1968 Speech Acts: an Essay in the Philosophy of Language. Cambridge, Cambridge University Press.

Sebeok, T.A. (Ed.)
1960 Style in Language. Cambridge, MA, MIT Press.

Seiffert, U. and Jain, L.C. (Ed.)
2002 Self-Organising neural Networks. Springer-Verlag, Germany.

Shu, N.C.
1988 Visual Programming. New York, NY, Van Nostrand Reinhold Company.

Tannenbaum, A.S.
1987 Computer Networks. Englewood Cliffs, NJ, Prentice Hall.

Teodorescu, H.N. and Jain, L.C. (Ed.)
2001 Intelligent Systems and Techniques in Rehabilitation Engineering. CRC Press USA.

Teodorescu, H.N., Jain, L.C. and Kandel, A. (Ed.)
2001 Hardware Implementation of Intelligent Systems. Springer-Verlag, Germany.

Teodorescu, H.N., Kandel, A., and Jain, L.C. (Ed.)
1998 Fuzzy and Neuro-fuzzy Systems in Medicine. CRC Press USA.
1999 Soft Computing Techniques in Human Related Science. CRC Press USA.

Tonfoni, G.
1991 Communicative cooperative interaction models: gapfilling processes by reformulation, in "Referate der Dritten Arbeitstagung Dialoganalyse III", edited by S.Stati, E.Weigand and F. Hundsnurscher, pp. 405-414, Max Niemeyer Verlag, Tubingen, Germany.
1994a CPP-TRS: on using visual cognitive symbols to enhance communication effectiveness, in "Proceedings of the Fourth International Symposium on Measurement and Control in Robotics: Topical Workshop on Virtual Reality," pp. 136-142, NASA, Lyndon Johnson Space Center, Houston, Texas, USA, December.

1994b "Writing as a Visual Art", (first English abridged version with J.Richardson, with a foreword by Marvin Minsky), Intellect, U.K. second version 2000, Intellect, U.K. and Scarecrow press U.S.

1995 A visual language for representing invisible aspects of natural language, in "Intelligent Tutoring Media", vol. 6, n°.2: 84-90, Intellect, Exeter, England, U.K.

1996a On visual text planning, processing and programming, in "Machine Learning and Perception", edited by G.Tascini, F.Esposito, V.Roberto and P.Zingaretti, pp. 99-106, Series in Machine Perception and Artificial Intelligence, Vol. 23, World Scientific, Singapore.

1996b Italy – An "inform-active" learning environment, in "Proceedings of the Round Table on Adult Learning and Technology in OECD Countries", 141-163, Organization for Economic Cooperation and Development, Philadelphia, Pennsylvania, USA.

1996c CPP-TRS for visual preprocessing, in "AI*IA Notizie", Anno IX, n° 2, 35-36, Rome, Italy.

1996d A visual news processing environment, in "Artificial Intelligence Review", vol. 10, n° 3-4, 33-54, Special Issue on "Integration of Natural Language and Vision Processing: Recent Advances", edited by P.Mc Kevitt, Kluwer Academic Publishers, Amsterdam, The Netherlands.

1996e High performance in text processing: the CPP-TRS environment, in "Working Notes of the AAAI- 96 Fall Symposium on Knowledge Representation Systems Based on Natural Language", 131-135, AAAI- 96 Fall Symposium Series, Massachusetts Institute of Technology, Cambridge, Massachusetts, USA.

1996f "Communication Patterns and Textual Forms", Intellect. U.K.

1997a CPP-TRS: a universal grammar for communicative interaction, in "Dialogue Analysis: Units, Relations and Strategies beyond the Sentence", edited by E.Weigand with E.Hauenherm, 233-249, Sonderdruck aus Beitrage zur Dialogforschung, Band 13, Max Niemeyer Verlag, Tubingen, Germany.

1997b A new paradigm for natural language processing and visual text planning the CPP-TRS model, in "Atti del Ventottesimo Congresso Internazionale della Società di Linguistica Italiana su Linguaggio e Cognizione", edited by C.Carapezza, D.Gambarara and Lo Piparo, pp. 419-434, Bulzoni, Rome, Italy.

1997c A visual interlanguage for iconizing documents, in "Actes du UNESCO Forum International des Sciences Humaines sur Le Plurilinguisme dans La Société de l'Information", pp. 50-65, UNESCO Publications, Paris, France.

1997d Qualitative reasoning in information processing and visual documentation management, in "Atti dell'Incontro dei Gruppi di Lavoro dell'Associazione Italiana per l'Intelligenza Artificiale su Apprendimento Automatico e Linguaggio Naturale", edited by L.Lesmo, pp.65-68, University of Turin, Italy.

1998a Cooperative documentation management, in "Proceedings of the International Colloquium on Multilingual Communication", edited by H.Bunt, University of Tilburg, The Netherlands.

1998b Intelligent control and monitoring of strategic documentation: a complex system for knowledge miniaturization and text iconization, in "Proceedings of the ISIC/CIRA/ISAS 98 Conference", pp.869-874, National Institute of Standards and Technology, US Department of Commerce, Gaithersburg, Maryland, USA.

1998c Relativity theory of information and communication in natural language, in "Journal of Artificial Intelligence and Society", vol.12, n°4: 322-327.

1999a On augmenting documentation reliability through communicative context transport, in the Proceedings of the 1999 Symposium on Document Image Understanding Technology, pp. 283-286, SDIUT 99, Annapolis, Maryland, US.

1999b Bridging the gap between Vision and Natural Language Processing; a visual metalanguage for supporting qualitative reasoning in textual design, in "Rassegna Italiana di Linguistica Applicata", Anno XXXII n°1, gennaio-aprile 2000, pp.3-36, Bulzoni, Roma.

2000a Visualization of textual structures, in "Iconic Communication", edited by M.Yazdani and P.Barker, Intellect, UK. 92-110.

2000b Model based reasoning upon information: CTML. a mark up language for tagging discourse and annotating documents, in "Sprache und Datenverarbeitung, International Journal for Language Data Processing", 24 Jahrgang, n.2, 2000, pp.59-68, IKS, Bonn, Germany.

2000c The making of specialised knowledge in information design through visual metaphors and spatial analogies, in "Studi Orientali eLinguistici", vol.VII, 2000, pp.235-255, Edizioni CLUEB, Bologna.

2000d Quality enhancing conceptual tools for medical decision making, in Future of Health and Technology, R.G.Bushko ed., Series in Health Technology and Informatics, pp.145-162, 2002, IOS Press, Amsterdam.

2000e Information transport through dialogue: a system for reducing fuzziness due to cultural and context shifts, in Proceedings of the Symposium of the International Association for Dialogue Analysis, M. Bondi, ed. June 15-17, 2000, Max Niemeyer Verlag, Tuebingen.

Tonfoni, G. (Ed.)

1998 "Information Design: the Knowledge Architect's Toolkit", Scarecrow Press, Lanham, MD and Intellect, U.K.

Tonfoni, G. and Jain, L.C.

2003a "The Art and Science of Documentation Management", Intellect. U.K.

2003b The Innovations in Decision Support Systems. Advanced Knowledge International.

Visualizing Document Processing. Mouton De Gruyter, Germany, in press.

Van Dijk, T.

1969 "Sémantique Structurale et Analyse Thématique", in Lingua, 23: 28-53.

Van Rooij, A., Jain, L.C., and Johnson, R.P.
1996 Neural Network Training Using Genetic Algorithms. World Scientific Publishing Company, Singapore.
Vonk, E., Jain, L.C., and Johnson, R.P.
1997 Automatic Generation of Neural Networks Architecture Using Evolutionary Computing. World Scientific Publishing Company, Singapore.
Wilensky, R.
1983 Planning and Understanding, Reading. MA, Addison Wesley.
Winston, P.H.
1992 Artificial Intelligence, Reading. MA, Addison Wesley.

Index